Off-Trail Adventures in Baja California

Books by Markes E. Johnson

Discovering the Geology of Baja California: Six Hikes on the Southern Gulf Coast (2002)

Atlas of Coastal Ecosystems in the Western Gulf of California (2009), coeditor with Jorge Ledesma-Vázquez

Available from the University of Arizona Press

Off-Trail Adventures in Baja California

Exploring Landscapes and Geology on Gulf Shores and Islands

Markes E. Johnson

Markes E. Johnson

THE UNIVERSITY OF ARIZONA PRESS
TUCSON

For Jorge Ledesma-Vázquez and David H. Backus,
who shared much of the journey.

The University of Arizona Press

www.uapress.arizona.edu

Library of Congress Cataloging-in-Publication Data
Johnson, Markes E.
Off-trail adventures in Baja California : exploring landscapes and geology on gulf shores and islands / Markes E. Johnson.
pages cm
Includes bibliographical references and index.
ISBN 978-0-8165-2130-2 (pbk. : alk. paper)
1. Geology—Mexico—Baja California (Peninsula)—Guidebooks. 2. Baja California (Mexico : Peninsula)—Guidebooks. I. Title.
QE203.B34J65 2014
557.2'2—dc23
2013034783

Manufactured in the United States of America on acid-free, archival-quality paper containing a minimum of 30% post-consumer waste and processed chlorine free.

19 18 17 16 15 14 6 5 4 3 2 1

Contents

Illustrations

Figures

Maps

Tables

Plates

Acknowledgments

ONE OR MORE RESEARCH TRIPS each year were made to Baja California from 1989 to 2013 from the Williams College Department of Geosciences. In total, 77 Williams students took advantage of January field courses or Spring Break excursions to study in the region. Among them, 15 signed on for extended honors theses or independent studies: Marshall Hayes ('92), Mark Mayall ('93), Hovey Clark ('94), Jen Zweibel ('94), Laura (Libbey) Blackmore ('95), Max Simian ('95), Jon Payne ('97), Patrick Russell ('97), Alison Kopelman ('97), Cordelia Ransom ('00), Mike Eros ('03), Paul Scudder ('05), Kristen Emhoff ('09), Dan Perez ('10), and Peter Tierney ('10). Their eyes became trusted extensions of my own. Most of this work was made possible by grants from the donors of the Petroleum Research Fund under the auspices of the American Chemical Society. In addition, some 80 Williams alumni or alumni spouses took part in college-sponsored tours to Baja California under my direction.

Leon Fichman and Ivette Granados (Loreto), as well as Tim Hatler (La Ventana), rendered outstanding services as outfitters in charge of island logistics. Enrique Baeza (Kiki) from Loreto, Jesus Salvador Calderon-Aviles (Nono) from La Ventana, and Ricardo Arce-Navarro from Bahía de los Angeles were the skilled boatmen who always brought us safely back to home port. Mexican government offices issuing permits that made research possible in national parks and biosphere reserves included the Area de Protección de Flora y Fauna Islas del Golfo de California, Comisión de Areas Naturales Protegidas (CANP); Bahía de Loreto Parque Nacional (BLPN); and the Instituto Nacional de Antropología e Historia (INAH) en Baja California Sur. In particular, Alfredo Zavala (CANP) and Everardo Mariano-Meléndez (BLPN) are thanked for their cooperation and interest in our research.

Those who agreed to read early drafts of individual chapters or, indeed, the entire manuscript included Richard C. Brusca, Norman Christie, Tim Hatler, Jeannine Perez, and Rafael Riosmena-Rodríguez. Their suggestions and encouragement were most welcome and contributed to improving the final product.

My companions in exploration and teaching are Jorge Ledesma-Vázquez (Universidad de Baja California) and David H. Backus (Williams College), to whom this book is gratefully dedicated. I owe heartfelt thanks to my spouse, Gudveig Baarli, and my son, Erlend Johnson, for granting me the freedom to spend so much time apart from them in Baja California, although they too were sometimes enlisted for field studies and more often for technical computer support. Their only recompense was that I always returned home with a sweeter disposition ready to resume my duties as a husband and father. Finally, the office of the Dean of Faculty at Williams College contributed a generous subsidy to the University of Arizona Press that helped make this book a reality.

Advice to the Reader

THIS BOOK DESCRIBES a sampling of places on islands in the Gulf of California and related shores of Mexico's Baja California peninsula. It is meant to celebrate spots where the intersection of geography, geology, and ecology meet to instill in the observer a deep understanding of place and time. We humans have many peculiarities that set us apart, but our innate curiosity is surely among our most definitive traits. We want to know what makes a thing go, how a system functions, and where we might fit into the bigger picture. Whole years may be devoted to reaching but a few "ah-ha" moments, when the clarity of a particular setting strikes a chord within us and we come to know something that occupies more space and meaning than found in our small, personal lives. I am never more alive than when one of those flashes of insight strikes. Months, maybe even years, of normal living stretch between such moments. To some extent, however, the special moments can be anticipated and actively cultivated.

For twenty-five years, I have returned on an annual basis (mostly during the month of January) to explore the wild landscapes and little-known geology of the Baja California peninsula. Foremost, I am a teacher. During most of the year, my work is confined to the classroom, where I try to interest college students in the story of our planet's development. During the summers, I am free to conduct field studies that have taken me to other destinations such as Western Australia, southern China, Inner Mongolia, Siberia, northern Norway, and parts of Canada. But at least once a year for several weeks at a time, I am lured back to Baja California like a moth drawn to the light. Not that the geology of other places is less exciting, but the atmosphere of Baja California is a deep-soothing balm that restores my physical and intellectual energy. Perhaps I find it so because students usually join me on my travels there. The lines between teaching and research become blurred. I revel in sharing new insights with them just as much as with my research colleagues. More to the point, seeing the place through the eyes of novitiates has a rejuvenating power.

In a previous book, *Discovering the Geology of Baja California* (Johnson 2002), I focused on a single region covering only 9.5 square miles (25 km^2) at Punta Chivato between the towns of Santa Rosalia and Mulegé in Baja California Sur. Punta Chivato pushes into the Gulf of California as a promontory near 27° north latitude (equivalent in latitude to Sarasota, Florida). My goal in developing that guidebook was to provide a menu of six hikes, each with increasing physical and mental demands, while demonstrating how the Baja California peninsula was assembled over many millions of years and how the Gulf of California came to be one of the planet's youngest seaways. Punta Chivato remains a wonderfully instructive place to visit.

This book differs only in its attempt to introduce the reader to nine additional localities spread between 24° and 30° north latitude. The north-south linear distance amounts to 385 miles (620 km). If the Punta Chivato promontory may be considered a kind of microcosm of the entire gulf region along the Baja California peninsula, as I argued earlier, then the addition of several more focal points along the length of those shores should enrich perspectives on the region and make the journey more evenly representative of the whole. The same geographic spread in Florida is equivalent to the north-south distance of that state including the Florida Keys.

The geographic shape of Baja California as a long and narrow territory lends itself to physical aloofness. The peninsula sits apart from the rest of the world, and its great desert sanctuaries are unable to support large population centers. This combination offers the opportunity for one of life's luxuries: the freedom to undertake an exploration (mental and physical) to the exclusion of all other considerations and outside distractions. Over my many visits to Baja California, I have watched as roadside trenches were excavated for the laying of fiber-optic cables from one end of the peninsula to the other. Now the place is wired so that even the smallest towns support Internet cafes. If you must stay connected to the wider world, you may do so. The great charm of the place, however, is that one can escape off-grid into the landscape and ignore the static noise that clutters our lives.

Like everyone else, I find myself challenged to maintain a healthy balance between family and career. While one is passing through midlife, the strain on any such balance only increases as responsibilities to aging parents and maturing children mount on opposite sides of the generational divide. My immediate and extended family understand that when I leave for Baja California, I will be incommunicado most of the time I am away. Our preferred camping spots are remote. Cell phones based in the USA do not function in Mexico. Except for the occasional fishing camp, most of the islands in the Gulf of California are uninhabited. When pressed to do so, I have reluctantly agreed to bring a satellite phone along as a safety precaution.

So it is that Baja California represents for me a physical retreat from an unruly world, but also a zealously guarded state of mind. Such a state of mind requires one reality only: that nearly every fiber of one's body remains intent on experiencing the panoply of life on land and in the adjacent sea and that every bit of intellect remains open to the web of remarkably persistent relationships having developed there through time. This development is clearly recorded in the rocks of the peninsula and its associated islands. The signs are everywhere to be found. The most important judgment to be made on rising in the morning is to predict how sunny or overcast the day might be and to decide what kind of clothing and equipment are most appropriate for the day's adventure. Nothing else matters much, except for the logistical details of food and water rations.

This book is about finding places prized for the stories they reveal about the land and how it relates to the environment through roots that thread back in geologic time. The armchair traveler has no option as a reader but to follow along to those places of my choosing. However, if you are inspired to abandon the congeniality of your armchair and stand where I have stood at the end of a hike illuminating a particular setting, then you have the resources necessary to explore your own pathways and seek your own interpretations hidden in landscapes that bring a kind of muscle-sore but well-earned satisfaction.

The chain of events that first brought me to Punta Chivato in 1991 was entirely fortuitous (Johnson 2002). I had no plan, but I was dropped into a landscape rich in reciprocal relationships of a geological and ecological nature. The discovery of a place where few geologists or biologists had trod before meant that my students benefited from opportunities to conduct original field studies that added to our understanding of the area's natural history. By the late 1990s, it began to dawn on me that satellite images might play a more deliberate role in tracking down similar spots worthy of more exploration and study. Eventually, this led me to visit some of the islands in the Gulf of California. Because of the difficult logistics involved, satellite images were indispensable in guiding our field parties to those places most promising as study sites. We simply could not afford the time and resources required to blunder into a good thing. Four islands, among several others so explored, are featured in this volume.

Although far from an absolute survey of the lands bordering the Gulf of California, enough had been traversed since 2002 for compilation of the *Atlas of Coastal Ecosystems in the Western Gulf of California* based on my experiences and those of my research colleagues (Johnson and Ledesma-Vázquez 2009). That volume focuses on fossil-rich limestone of Pliocene and Pleistocene age and modern carbonate environments. The marine invertebrates and algae that precipitate limestone are found living in relatively shallow waters adjacent to the geological deposits of a comparable nature, now found abandoned by the sea on land. The prize hidden within the covers of our academic volume is a computer disk with a set of 26 overlapping satellite images. Stretching from the delta of the Colorado River in the north to the famous arches of Cabo San Lucas at the terminus of the peninsula in the south, these images provide nearly 100 percent coverage of the coastal zone along the gulf shores of Baja California and its related islands. Given the technology widely available today (including Google Earth, which offers worldwide coverage but only in the visible spectrum), anyone who has access to a computer terminal can check out a place before taking the trouble to actually go there.

Were unlimited financial resources at my disposal, I would not hesitate to make a comprehensive tour of each island and every stretch of the related gulf coast by small aircraft. Through satellite reconnaissance, however, much the same task can be accomplished at a computer terminal. Once a promising locality is identified, it is always possible to take advantage of other viewing opportunities, such as flights on regularly scheduled airline routes that bring travelers back and forth along the peninsula's coastal axis. Even at a cruising altitude of 30,000 feet (9,000 m), quite a lot can be seen under the generally clear skies of the gulf region. Many of the photos in the color section of this book were taken on flights with commercial carriers, and the information so gathered was very useful as a supplement to satellite images in refining plans for our island visits. In the end, however, there is no better substitute for knowing a place than an on-the-ground immersion.

I hope this book will help stimulate ecotourism in Baja California, still one of the largely untouched but easily reached wilderness areas in the world. Off-road camping is unhampered by fees or regulations in most places. Smaller islands close to the peninsular mainland are more accessible than others, but payment of park fees through licensed boatmen is proscribed for protected areas. Mexican authorities in charge of biosphere reserves control all larger islands, where visitation inland from beach areas requires a research permit. Although this book profiles some individuals who prefer to make their explorations solo, it is strongly advised to travel in the company of others and to make sure that someone outside your group knows where you are going and how long you plan to be there. The winds, tides, and currents in the Gulf of California can be treacherous, and using the services of local fishermen and boatmen is a wise precaution in exploring remote coasts and distant islands. Mexican laws prohibit the removal of archaeological artifacts, fossils, and biological products. The proper ethic is to take as many photographs as you would like and leave behind only footprints.

Author's note: All place names used throughout this volume conform to the 2009 edition of the *Baja California Almanac* (Las Vegas, NV: Baja Almanac Publishers).

Off-Trail Adventures in Baja California

1

Cataviña

Gateway to a Peninsular Wilderness

> The Space Wanderer was finding his adventures so satisfying and stimulating, so splendidly staged, that he was shy about questions—was afraid that asking questions might make him seem ungrateful.
>
> —*Kurt Vonnegut Jr.*, The Sirens of Titan

EXPERIENCING THE HIGH DESERT around Cataviña is like reaching an abrupt warp in the time-space continuum. The rocks, topography, and native vegetation all conspire to bring you to a place of such singular contrasts that the tableau might be imagined to exist on another planet outside our solar system. No spacecraft for intergalactic travel is necessary. Access is by Mexican Federal Highway 1 from El Rosario on the Pacific coast (see map 1, locality 1), a distance of only about 65 miles (105 km). The distinctive Cataviña Boulder Field comprises the northerly section of protected lands belonging to the Baja California Central Desert Natural Area, poised on the backbone of the peninsula about 180 miles (290 km) due south of the USA-Mexican border. After the sign at the junction for Faro San José, there is another turnoff on the south side of the highway leading to an ample parking area surrounded by massive granite boulders, still a short distance west of the unremarkable village of Cataviña. I have lost count of the number of times I have camped overnight in these surroundings with student groups on our journeys southward

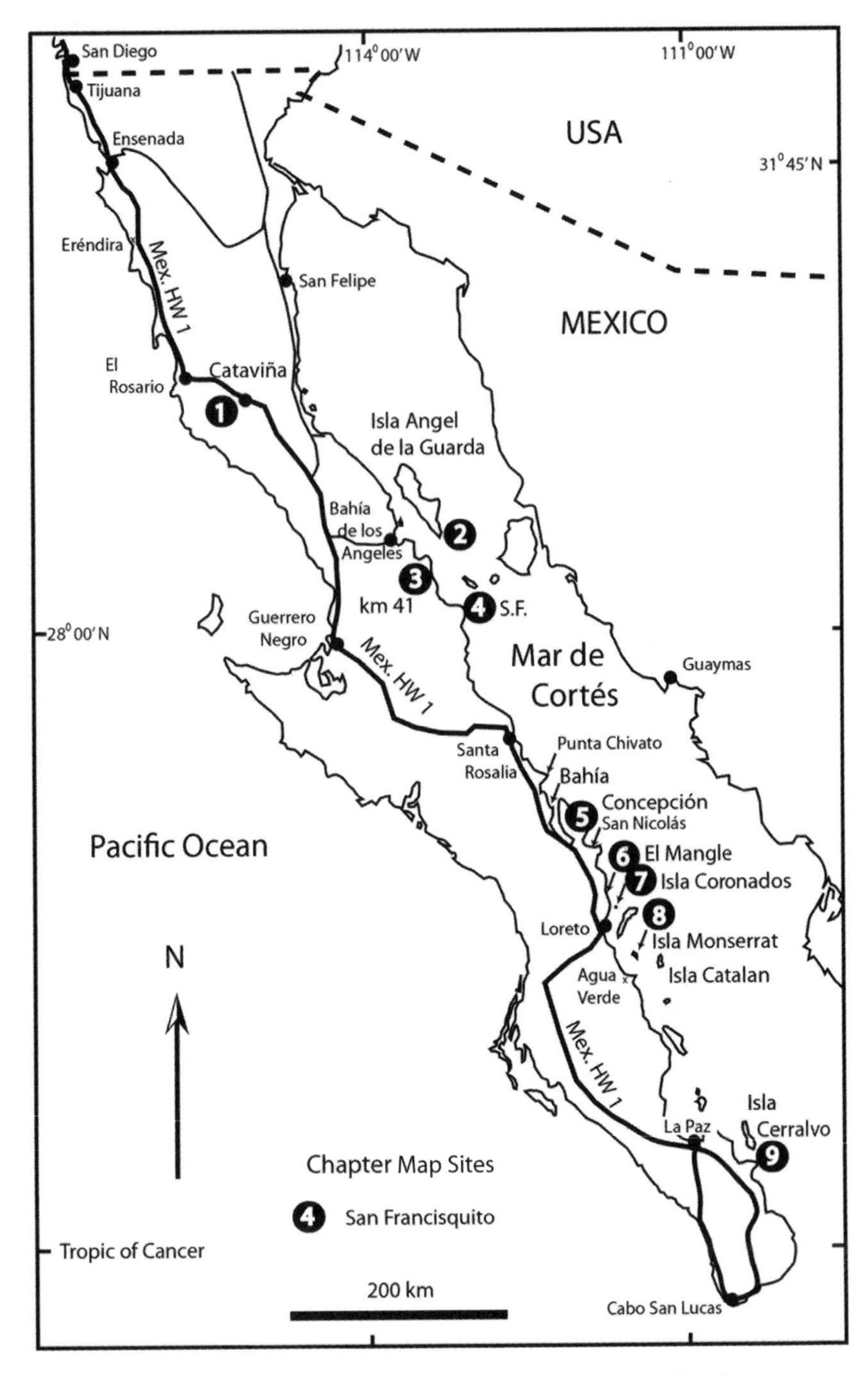

Map 1. Localities in Baja California and associated islands. Map by author.

to the Gulf of California. It is one of my favorite camping spots, with rites and rituals uniquely befitting a landscape that embodies both a physical and a cerebral gateway from one state of being to another.

The weather was still comfortably warm during the late afternoon on a January day in 1990, when I first piloted a van to the parking area safely apart from the harrowing confines of the shoulderless highway. The scene, on alighting from the van, became permanently etched in my memory. It was one of those rare days in the lunar cycle when an early moonrise over the eastern horizon coincided with the setting of a hazy sun in the west. Fine dust suspended in the air created an atmospheric illusion of a full moon, much magnified in size. The place had an entirely extraterrestrial feel to it, as if located on a planet with not one but two moons held on mesmerizing paths in close orbits. Almost in a trance, we moved from the vehicles to one of the great boulders on the edge of the parking area, and we clambered to the top for a panoramic view of the countryside. Everywhere, as far as the eye could see, were rounded bodies of pink granite that ranged in volume from the equivalent of a small house to a spacious kitchen to a generous bathtub. In places, the boulders are piled in great heaps, like so many enormous beanbags cast aside by a playful child of gargantuan size and strength. The native race of giants has vanished. A heavy silence lay over the land, interrupted only by the noise of a passing car or truck—not so common late in the day.

The true giants of the land are the massive cardón (*Pachycereus pringlei*), commonly found through much of the Sonoran Desert with relatives such as the saguaro (*Carnegiea gigantea*) ranging far to the north in southern Arizona (Roberts 1989). Here, however, their size is extraordinary: as tall as 65 feet (20 m).[1] A raft of more-diminutive cholla species (genus *Opuntia*) populate the landscape, with some, such as the silver cholla (*O. echinocarpa*), endemic to the region. All are exceedingly thorny, with padded green flesh showing stronger or weaker traces of a diamond-shaped lattice in the supporting woody skeleton beneath. The cholla cacti are restive natives that jump to attack anyone or anything that innocently approaches. Distal segments on cactus branches detach all too easily from the rest of the plant and cling with hooked spines to clothing, hide, or even the tough soles of field boots. The propagating segments are difficult to dislodge.

By far, the most peculiar resident is the boojum tree, or cirio (*Fouquieria columnaris*). It represents an odd species grouped with the spindly ocotillo (*F. splendens*) in the same lonely genus. The lean taxonomic assignment erected by botanists gives some notion of the wildly endemic character of this orphaned plant lacking few close relations. If trees were to grow on Mars, they surely would be found to denote a distant relative in the genus *Fouquieria*. Tagged in Spanish as a *cirio*, the word calls to mind the image of a long and tapered candle, stout at the base but progressively slender to a vanishing point at the top. This works as a model, but the body of the candle is festooned with the lacy growth of pencil-short branches bearing tiny green leaves during wet years. A fingerlike splay of tubular blooms crowns the towering distal tips of the plant, attracting hummingbirds and bugs that facilitate cross-pollination. More often than not, the adult tree is pole-like, but individual plants sometimes fork into multiple branches that may curve out in bizarre directions. These odd denizens of the high desert can reach heights of 60 feet (18 m).[2] No one knows for sure, but the maximum age of individual plants may surpass 250 years.[3]

Another typical resident is the barrel or compass cactus (*Ferocactus acanthodes*), so named because of its propensity to lean toward the south or southwest. It is thought that by pointing in this direction toward the most intense summer light, the compass cactus minimizes the most harmful effects of the sun.[4] The one inhabitant I most closely identify with is the old man cactus (*Lophocereus schottii*), whose upper columnar branches are bewhiskered with a growth of long, gray spines that provoke an anthropomorphic sense of sage-like respect. The various species mentioned in this all-too-brief account can be found in the color photo taken in January 2005 at the southern edge of the Cataviña Boulder Field (see plate 1) during an El Niño year of plentiful moisture that left the landscape as green as I have ever seen it in more than two decades of travels through Baja California. In the distance below the mountains on the horizon are the waters of Laguna Chapala, normally a dry lakebed.

As a geologist/paleontologist, I confess that my fascination with these desert plants is related to the surprising array of internal woody frameworks that serve as rigid skeletons. By and large, most of the organisms that become fossilized and preserved in the geological record require hard parts and some means of relatively swift burial.

Although my specialization is with marine invertebrate fossils, I salivate at the thought of discovering a fossil cactus. If such a thing might be found, I suppose the imprints of its internal, woody skeleton should give it away. However, desert wood has a more practical reason to be of keen interest to the camper. After allowing my student charges sufficient time to relish the Cataviña atmosphere and to secure their tent sites, I send them out to harvest a supply of firewood from the skeletal remains of fallen cacti.

The most memorable evening I spent camping at the turnoff outside Cataviña was in late January 1996, when my companions that season were two thesis students, Jonathan Payne and Alison Kopelman. We had been conducting fieldwork on the Pacific shores of the peninsula around El Rosario and farther to the north near Eréndira, where strata of Cretaceous age are well exposed. In fact, both students had discovered fossil wood entombed in the rock layers that constituted their research projects. Jon's thesis, later to be published, included a description of limbs from some kind of evergreen tree that became mixed into a coastal marine conglomerate.[5] Alison's study entailed even larger conifer tree trunks buried in river sediments. Given the success of my students, no wonder I have dreamed of someday finding fossil cactus wood. We were on our way to Bahía de los Angeles for no other reason than to give the students a taste for the opposite coastline on the Gulf of California, before returning home.

That particular evening, however, was memorable for the audience we drew to our campfire. Dinner was cooked over a bed of coals and involved the preparation of some beef as a filling with sautéed onions and green peppers rolled into pan-warmed tortillas. After the dinner dishes were washed and put away, we added more desert wood to the fire to drive away the night chill. Cactus wood from some species is surprisingly dense and burns very well. The resins found in certain woods, especially the hollow-lattice skeletons of the cholla, give off intense heat and provide colorful firelight. It is hypnotizing to watch the flames that suddenly issue forth in colors of metallic-copper green and blue, while the embers decline to a steady glow. The low but strong light provided by our fire illuminated the sides of the largest granite boulders outside our huddled circle. In the distance came the plaintive cry of coyotes, and before we knew it, shadows cast against the stony walls included those of two circling animals with flowing tails caught in the flickering firelight. The smell of our dinner was more than they could abide, but they were shy guests and soon left.

I have a personal boulder where I usually make my bed in the open at the Cataviña campsite. My rock is well beyond the campfire and elevated enough to give me a kind of perch above the immediate surrounding area. With the campfire doused, the night chill of the desert air comes swiftly, and a warm sleeping bag is a welcome comfort. The stars shine with an intensity that suggests one might reach out and collect them at arm's length.

My colleague, Jorge Ledesma, enjoys repeating the story from his student years when he set out from Mexico City late in the day, headed for a conference in the coastal city of Acapulco some 200 miles (320 km) to the south. Night fell as his party climbed out of the natural basin in which the great city sits with all its smog. They stopped the car at the mountain pass to breathe the fresh night air and soothe nerves jangled from the press of traffic on their exit from the city. Not entirely in jest, someone in the group asked, "What are all those bright lights up in the sky?" With some alarm, it is possible to suppose that children might grow up in a dense urban setting and spend the rest of their natural lives where real pollution combined with light pollution render it all but impossible to see the stars.

Generally, it is during the first week of January when I arrive in Cataviña on my annual tour. Such is the tumult of final exams and the ensuing holidays that I wait to make any New Year's resolutions until this very moment on my personal rock. It is my custom to review the events of the past year and my expectations for the year ahead during just this pause before a fitful slumber. Many hopes and some grievances cross my mind in a disorderly stream of consciousness. The world of academia can be treacherous, and not without its petty rivalries. Some of my nonscience colleagues are suspicious about my motives for leaving the campus this time of the year during the worst of the winter weather. My escapes to the southlands are viewed as pleasant holidays. Others among the scientists wonder when I will quit Baja California and move on to other worthy pursuits. They do not understand the size of the place and its unending possibilities for studies that pull in many directions.

More knowledgeable about the areas of geology in which I work are colleagues who render peer reviews on papers submitted for publication in scientific journals. I am grateful to the vast majority, who keep me from making embarrassing mistakes and oversights. A few others are the source of much frustration. Two personalities reveal themselves in the exercise of this vital task: those who look for ways to help strengthen a study and those who seek every opportunity to destroy that which challenges their perceived mastery of a discipline. The tangible handiwork of my profession consists of the papers I manage to see published, many with student coauthors. Essentially, I am an academic stonemason, who steadily adds one stone after another in an effort to create a kind of architecture in honor of the natural world. Some stones require more craftsmanship than others to achieve an acceptable fit. Each stone successfully placed during the past year and each that remains a work in progress is revisited in my mind on this night.

I sleep and awake intermittently to find a more comfortable position on my rocky perch. Each time I turn, I observe the constellation Orion shifting to a new place in the night sky. The great hunter moves ever onward in eternal pursuit of his prey. As the night wears on, my reflections gradually shift from what has been to what will be. My thoughts turn from the people and issues that influence my life at home to those here in Baja California with whom I interact on an annual basis. It is a varied and colorful set of "locals" who contribute in no small way to the advances in our studies. I look forward to their fellowship every year, and they greet me as if I am returned to my rightful home. Among them are expatriate Americans, such as Marge and Jere Summers, living full time at Punta Chivato. The Baja California peninsula has attracted expatriates from other lands, as well. I relish my conversations with the Brazilian Leon Fichman in Loreto, whose knowledge of every far corner in the Gulf of California brings a nuanced overview of the place. His reliable boatmen have seen to our successful explorations of many such out-of-the-way places, always bringing us safely back.

Foremost, I treasure the connections with Mexican families introduced to me through my colleague Jorge. Perhaps tomorrow we will be on our way to the little village spread along Bahía San Nicolás, tucked below the outer shores of Peninsula Concepción (map 1). I think about the patriarch of the extended family, don Chico. He never fails to welcome us with an air of earnest dignity, inviting us to sit for coffee in the chairs around doña Francisca's open kitchen, sturdy seats carved by his own hands from the trunks of fallen date palms. It is here that I witness the old ways, tinged by the intervention of the new. Near the gate to don Chico's family compound are the decayed remains of a white-plastered Nayarit canoe,[6] fashioned from a single great log acquired by his father somewhere on the mainland coast of Mexico near Mazatlán. No trees of the girth required to make such a canoe grow anywhere on the peninsula. The value of such an imported watercraft was paramount to the family's survival. Each of don Chico's several sons now has his own fiberglass *panga*, powered by an outboard motor. Don Chico recalls the hard toil of fishing from the canoe as a boy with his father. They often went to sea for several days at time. The canoe is heavy, but it could be put under sail when the winds were favorable.

There are no electric power lines leading to San Nicolás. Instead, the federal government made it possible for families to acquire at minor cost a system of solar panels affixed to simple pipe stands. One of the contraptions stands at a corner of the open kitchen, and it can be hooked up to a small television set. Don Chico smiles every so slightly and tells us about a neighbor man who asked the government to repossess his solar panel only a year after the installation. It seems that his spouse quickly became addicted to the daily soap operas aired from Mexico City. The household rapidly fell into disorder. Duties were left unattended. The man's wife was abducted by the serial fictions of other families, now more important to her. Something had to be done to restore the former order of things. I come to Baja California again and again not as a Luddite but as a pilgrim wishing to experience the place in its purest form of nature without any complications from the outside world.

The grounds around the antique Nayarit canoe on don Chico's property are cluttered with piles of rocks, mineral samples, seashells, and fossils of museum quality. He has gathered these things through a lifetime of experience. He remembers where each item came from, and he knows how to answer our questions in a way that gives added meaning to each object. There is an air of nobility about this man, who is assured in his sovereignty. He knows the place and generously leads us to cliff faces, where the fossils are exposed, or to abandoned mine pits, where chunks of manganite ore [MnO(OH)] are still plentiful. In addition, he is a walking encyclopedia regarding the region's plant lore for medicinal and other practical uses.

A rosy glow intensifies in the eastern sky above Cataviña. At last, the night's long interlude of internal conversations is coming to an end. I prop myself up on an elbow and look out from my cradle on the rock. Several of my students are awake, sitting draped in sleeping bags and cross-legged on some of the boulders around me. They wait for the sunrise in silent meditation. A new day is about to begin. Through the night, I have undergone a ritual that separates me from all the worries and distractions of my home life. I am inured, again, to the natural rhythms of a place with endless attractions. We are about to indulge in one of life's greatest luxuries. We will spend the next few weeks exploring new places unfettered by outside concerns. I feel as though I am reduced to a simple water molecule absorbed by osmosis through the semipermeable outer membrane of a giant cell. This amazing cell contains many organelles representing all the varied ecosystems in a peninsular wilderness on the sea. Our progress cycling through those ecosystems is all that matters.

The fire is reignited, and a kettle of water is put on to boil. Students scurry here and there packing sleeping bags and tents, or attending to their morning toilette. Soon, the vans will be packed and ready to go. But before we rejoin Mexican Federal Highway 1 on our travels south, we take an hour to circle around the campsite in a wide arc and consider the abundant natural riches.

FEATURE EVENT

A Short Hike in the Cataviña Boulder Field

Large and small, where did all the granite boulders come from? Apparent anywhere we stand, the question is an obvious place to start. The boulders are not roughly broken and did not result from some cataclysmic collapse of the Earth's crust. Regardless of size, all are globular in shape and rather rotund in profile (see plate 1). How they came to be piled together in great heaps is hard to imagine. The means by which the boulders were eroded from a solid granite mass is hard to visualize, because we see before us something closer to the end result of a long process. Spheroidal weathering provides the answer: physical and chemical erosion under normal atmospheric conditions advance from all directions in rocks naturally fractured by joints that define equilateral cubes. Where the jointing inscribes really big blocks, the process yields huge boulders. Where the jointing is on a smaller scale, the process results in lesser boulders. A daily rhythm of hot days and cold nights in an arid climate that rarely experiences much moisture promotes the exfoliation of thin granite sheets along the exposed margins of fractures from all sides. The material thus shed by this process continues to break down on the desert floor. Everywhere we pause, the coarse desert soil is composed of loose quartz and plagioclase crystals with finer constituents provided by the disintegration of micaceous minerals such as biotite. The name applied to this sort of very immature soil derived from granite is *grus*. Because of the extensive production of the material, tent stakes are easily driven into the ground here.

I am an unapologetic disciple of the nineteenth-century geologist Charles Lyell,[7] who put geological education on a solid footing with the many editions of his famous textbook, still available in facsimile editions. No matter how many times it was revised, his influential book featured the same frontispiece illustrating the Roman Temple of Jupiter at Pozzuli near Naples, Italy. Few columns of the original temple remain standing today, but they are indelibly marked by borings made by bivalves capable of dissolving a small living chamber in the marble. Clearly, the bivalves had no access to the temple when it was built on dry land. But when the temple was flooded as the Bay of Naples went through episodes of subsidence, the bivalves arrived and did their work. The highest matching levels on three standing columns where the borings end show the maximum height to which the Tyrrhenian Sea rose. Today, the temple floor stands dry once more. The object of the exercise was for Lyell to demonstrate how local changes in sea level are natural events that transpired during historical time.

My most acute Lyellian insight came while visiting an old section of the ramparts on the grounds of the Imperial Palace in Tokyo, Japan. Much of the original fortification was destroyed during World War Two, but the barracks known as the Hyakunin Basho remain intact flanking some of the inner castle walls. In these barracks were posted the one hundred samurai warriors who stood guard over the castle, the oldest parts of which date back to 1590.[8] The ramparts near the barracks were constructed from large blocks of igneous rock, hewn in rectangular pieces approaching 5 feet (1.5 m) in the largest dimension. Many blocks in the wall show unmistakable evidence of spheroidal weathering after the passage of four centuries. Some blocks, including those without much decay along the seams, show etchings of a clearly circular pattern (figure 1). Given enough time and the absence of any repairs, the walls of the castle are doomed to crumble into a pile of large boulders.

The seams along which the quarried stones in the Tokyo castle were so carefully fitted by master stonemasons are the equivalent of natural fractures that formed in solid Cretaceous granite exposed in the Peninsular Ranges along the central spine of Baja California. The ongoing

Figure 1. A single block in a rampart wall on the grounds of the Imperial Palace, Tokyo, showing spheroidal weathering. Photo by author.

erosion that attacks these joints is entirely natural, and it occurs with deleterious results on a time scale measured in the mere centuries over which the Imperial Palace has stood at the heart of the capital city. Mexican Federal Highway 1 enters the Cataviña Boulder Field from the northwest at an elevation over 1,950 feet (600 m) above sea level. This is one of the lowest regions where the highway could be threaded from one side of the peninsula to the other. The same granite rocks are widely exposed throughout the upper half of the peninsula all the way north of the border around San Diego.[9] Exposures are less continuous through the southern part of the peninsula but also occur in the mountains west of Loreto and south of La Paz. The famous sea arch at Cabo San Lucas is eroded by waves impacting much the same Cretaceous granite. In effect, the Cataviña Boulder Field represents a breach in the mighty wall of the Peninsular Ranges dividing the Gulf of California from the Pacific Ocean. The wall has stood for millions of years and remains quite high to the north. It, too, is doomed to ruin over the long march of geologic time.

Prior to exposure in the Peninsular Ranges, where did all the granite in the great wall originate? This is the next most logical question that begs for an answer. Granite is an intrusive igneous rock, which means that it is the product of magma that never quite reaches the Earth's surface. The large grains of silica and plagioclase, together with smaller bits of biotite that are the textural constituents of granite, precipitate from molten rock that cools slowly over time while insulated by a thick cover of overlying rocks in the Earth's crust. Essentially, the granitic melt is held in a magma chamber far below the surface of the Earth. Individual chambers may be connected to the surface by volcanic necks, but any magma that escapes to the surface will not result in granite. Surface flows from such a volcano may be similar to granite in chemistry, but the flowing lava is subject to rapid cooling that produces a finely crystalline rock called rhyolite. This rock is found on the outskirts of La Paz. I have encountered rhyolite at Punta Mercenarios and the adjacent Ensenada San Basilio north of Loreto.

Because we find few if any rhyolitic rocks in the central Peninsular Ranges, however, we might assume that the magma chambers were not well connected to the surface or that all the evidence for possible connections was removed by erosion. An old magma chamber in which the melt solidifies in place is called a pluton. Granite plutons now exposed in mountain ranges were once buried far below the surface of the Earth at depths ranging from less than 1 mile (1.6 km) to more than 7.5 miles (12 km) deep. Drilling in the Three Virgins volcanic region, for example, encountered Cretaceous granite basement at a depth slightly more than 1 mile (1,700 m).[10] This indicates the emplacement of an epizonal pluton of comparatively shallow origins. The vestiges of former plutons in the Cataviña Boulder Field are likely to have been buried at a similar depth below the surface. When the rocks above a buried pluton are removed by erosion, the granite is said to be unroofed.

Before erosion goes too far, as it has in the Cataviña Boulder Field, surface exposures of the massive granite can be readily mapped from the top of an unroofed pluton. If a given pluton exceeds 40 square miles (103 km^2) in area, it is classified as a batholith. The 1971 geological map for the state of Baja California (northern half of the peninsula) by Gordon Gastil and his associates shows many crudely circular structures exposed throughout the Peninsular Ranges that qualify as batholiths.[11] Elsewhere, the western margin of North America is packed with batholiths that attest to subduction of oceanic crust in trenches leading below the continent's active margin. Wedged apart by seafloor spreading at ocean ridges and partly pulled-down trenches under its own mass, oceanic crust is eventually melted and fed under enormous pressure to ready magma chambers (figure 2).

The grandiloquent Half Dome in California's Yosemite National Park represents an example of this end process. The unroofed half-dome profile lends itself wonderfully to an appreciation for the magnitude of a large granitic pluton, this one of an older Jurassic age and related to a mountain-building episode called the Nevadan Orogeny. Perhaps the world's most outstanding example is the Coast Range batholith of British Columbia, which is of comparable Cretaceous age to the plutons of Baja California. The Canadian colossus is up to 100 miles (160 km) in width and stretches more than 1,000 miles (1,600 km) in length. Ovoid structures within the megabatholith are described as having formed at depths of 4.35 to 7.5 miles (7–12 km) below the surface (Brown and McClelland 2000).

Coeval events in the development of plutons along the Pacific margin of Mexico before the opening of the Gulf of California were clearly less imposing, but they signify much the same tectonic processes

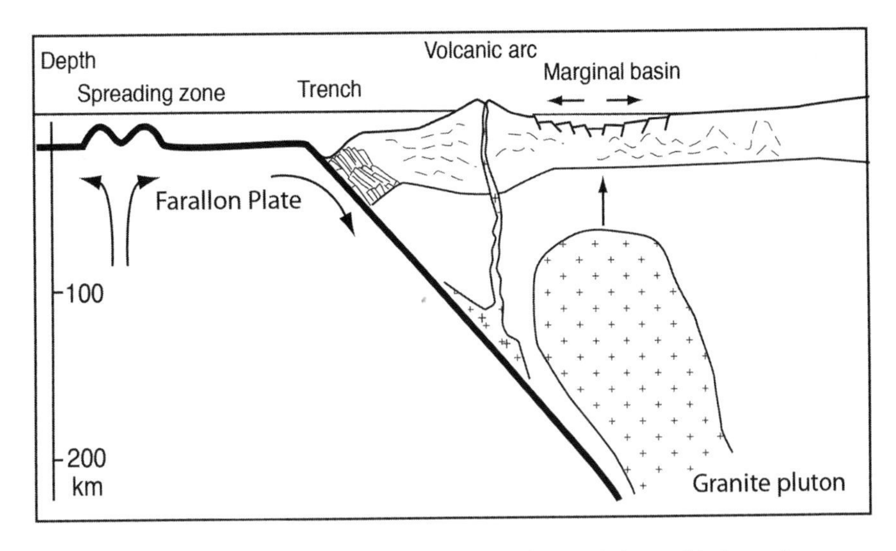

Figure 2. Cross section of the Farallon Plate subducted below the western margin of North America to produce volcanic arcs and granite plutons during the Cretaceous Period more than 80 million years ago. Original drawing by author.

related to the recycling of oceanic crust derived from tectonic plates such as the now-consumed Farallon Plate.[12] The present-day Gulf of California is the result of complex interactions between the spreading center of the East Pacific Rise and the North American mainland. The gulf began to open as a result of crustal extension over western Mexico as parts of the Farallon Plate were consumed and the leading edge of the East Pacific Rise collided with North America. This appears to have happened between 13 and 3.5 million years ago and resulted in the stretching and rupture of continental crust to form many of the present island highs and channeled lows in the Gulf of California (Ledesma-Vázquez et al. 2009). About 3.5 million years ago, the regional tectonic regime changed to accommodate movement along lateral fractures introduced by surviving segments of the East Pacific Rise captured within the gulf.

For now, it is enough to look south from our vantage spot on the Cataviña Boulder Field and see a nearby mesa that rises on the perimeter. A mesa is a flat-topped tableland, and the structure within sight is capped by dark basalt that represents extensive lava flows of

Pliocene to Miocene age ranging from approximately 3 to 15-plus million years old. Looking around to recognize other mesas on the far horizon, we can tell that cap rocks once covered the entire area and were stripped away to reveal the Cretaceous granite below. In other words, we can deduce from the physical arrangement of the eroded boulders, surrounded by remnants of formerly connected tablelands, that the core of the Cataviña pluton was unroofed relatively recently and continues to be laid bare by the ongoing attrition of the tablelands.

As we meander among granite boulders of all sizes released from captivity by spheroidal weathering, we encounter the skeletons of many different cacti that can be distinguished by the shape and construction of their desert wood. It is revealing to examine the skeletal remains side by side with healthy plants of the same species. The giant cardón, for example, has a stout trunk that forks into multiple branches of considerable girth. This girth does not conform to a simple circle, like a smooth belt drawn around the midsection, but takes on an accordion-like pattern with creased folds standing out from wedged indentions. A given branch may have up to a dozen vertical ribs that are located below the folds. The ribs are quite long and resemble the shape of wooden laths that might be purchased at a lumberyard. These natural laths exhibit densely packed fibers and are very strong. Considerable force is required to break them into lengths appropriate as firewood. In fact, the cardón ribs make fine walking staffs.

Similar in design to the cardón but consisting of flimsy sticks having a circular cross section on a simple rod, the internal supports of the bearded old man cactus are made of softer wood. The sticks are easy to break and burn rapidly. Each branch, which divides off from the main trunk of the cactus near the ground, contains about half a dozen woody rods. As with the cardón, these are embedded below the creased vertical folds that run along the entire length of each branch. When we detour around a large boulder, we come to a clearing where a half-decayed old man cactus stands with the green flesh of two limbs stripped clean away. The cactus looks like an amputee, but with naked bones protruding from the stumps of its arms. When we forage for desert wood to feed our campfires, we collect only the skeletal remains of cacti that have toppled over and decayed beyond the point where any photosynthesizing flesh remains intact to do its job.

Desert plants belonging to the cholla tribe possess the most distinctive of all woody parts. Branches are held erect by hollow, woody tubes from an inch and a half to four inches (4–8 cm) in diameter. These tubes are perforated by diamond-shaped holes that give the supporting structure a mesh-like look. The wood is hard and comparatively resistant to decay. If ever desert wood were to be found in fossil form, the cholla would be the most likely candidate to provide foolproof testimony to the fact. The plant family to which all cacti belong, the Cactaceae, is essentially native to the Americas, and the origins and rapid development of this highly successful family are difficult to trace. Thus, it would be most illuminating to discover a trove of fossil cacti. Although claims are sometimes made as to the discovery of fossil cactus material, none has proven legitimate (Rowley 1978). The fundamental problem is that cacti thrive in arid climates, where water-transported sediments are unlikely to entomb them. While fossil pollen and other plant macrofossils representing a wide range of trees and shrubs are found preserved in pack-rat nests from the Cataviña area (Sankey et al. 2001), the only cactus pollen recovered is identified as belonging to another barrel cactus (*Ferocactus* cf. *cylindraceus*). In the scheme of things, these remains barely qualify as fossils from the Late Holocene (i.e., only 1,770 years old).

If we assume that rapid burial of woody debris swept together in a rare desert torrent might occur, the main problem to overcome is the general porosity of desert regolith, or primitive soil, characterized here at our feet by the abundant grus derived from dismantled granite. The burial must prevent oxidation, which will occur in loose sediments. It is a problem that I sometimes ponder during long nights while camping in Baja California. One solution to all the difficulties of desert fossilization may be found in the thermal springs that abound throughout the region. El Marmol, outside the Cataviña Boulder Field only 12 miles (20 km) to the northwest, is the site of a former onyx mine, where the material left by mineral springs supported a commercially viable operation for many years.

Active mineral springs still flow at El Volcán in a valley north of the mine site, where lime-rich waters feed cone-shaped deposits of travertine. At various times, we have visited El Volcán in search of plant

Figure 3. Travertine deposit at El Volcán near the Cataviña Boulder Field, showing contemporary wood from the creosote bush as well as likely material from the borage family (*arrow*) undergoing fossilization. Photo by author.

remains incorporated in the travertine. There is striking evidence to support this rapid process of fossilization (figure 3), wherein a branch is half entombed by the mineralizing brew but its woody cortex with recognizable growth rings still protrudes in the air untouched. Plants caught in this process at El Volcán appear mostly to be stems from the Mormon tea plant (*Ephedra* sp.) and occasional branches torn from the creosote bush (*Larrea tridentata*), an evergreen that clearly exhibits growth rings. Woody remains likely to belong to shrubs (*Cordia* sp.) in the borage family also are present.[13] Technically, these examples fail to qualify as fossils, because the mineralization process is happening right before our eyes. However, this setting is another triumph of Lyellian observation. It means that the same basic processes of nature that can be witnessed under direct observation must also have been responsible for similar phenomena during the distant past. The trick is to locate ancient mineral springs preserved in the landscapes of Baja California that incorporated fossil plant debris dating back a few millions of years.

The morning is young, and the open road invites us to continue our journey. One thing is certain. Mexican Federal Highway 1 will lead us from the great plutonic backbone of the narrow peninsula to the Gulf of California.

Lesser imponderables cross my mind as I go about my livelihood. I understand how fortunate I am to have found my way here. Each year, I wonder which students under my brief tutelage will succumb to the spell cast by this place and take up a research career in geology. I remain awestruck by the native acuity of residents such as don Chico, who is well attuned to the interlocking environments around him. It shows that we need only be open to our surroundings to enjoy the wealth of nature at a high level of understanding, while our lives may be more fully absorbed with the basic business of living. The ambience of this country is powerful, but a similar ambience is not unknown in other places. What innate kernel of sensitivity sits within our makeup, urging us to seek out the stories usually kept hidden within the Earth? My life circumstances have been charitable to encourage this urgency within me. My peasant forebears in Scandinavia had few such opportunities, I think. Somehow, though, I imagine that a previous Markus Johansson may have been a stonemason, using explosives to quarry stone for the building of bridges, canal locks, and perhaps even parish churches, where he would pray for the protection of Barbara, patron saint of miners and quarrymen. And what might have been, were my ancestral line to have emerged from Asia? One of my ancestors surely would have been a master stonemason, quarrying and fitting the great stones to raise mighty castle walls in Japan.

Those builders who worked with stone from earliest times encountered marvels that awakened in them a wonder for the Earth. They had to learn to overcome any shyness about asking questions. Today, I am a devotee of Charles Lyell, and I practice the kind of "actualism" that informs our questions about the Earth. That is to say, the actual physical mechanics and life forces in the real world directly in front of us provide ample clues to reach an understanding of the deep geological past. And I would gladly follow the precepts of Lyell as a Space Wanderer to the shores of the Winston Sea on Titan, the distant moon of Saturn, where natural wonders undreamed of in our cosmogony still wait to be realized.

2

Investigations on a Guardian Angel

The difficulties of exploration of the island might be very great, but there is a drawing power about its very forbidding aspect—a Golden Fleece.

—*John Steinbeck,* The Log from the Sea of Cortez

WHAT POSSIBLE TREASURES draw a sane person over unruly waters to risk investigation of an island infested with serpents and lacking the rudiments of decent shelter or fresh water? Some of those who approached or made safe landfall on one of the most imposing islands in the Gulf of California left varying impressions. Discovered in 1539 by the mariner Francisco de Ulloa sailing under orders from Hernán Cortés to explore Mexico's western shores, Angel de la Guarda is the second largest of the gulf's islands (see map 1, locality 2). Roughly 42 miles (67.6 km) long by 6 miles (9.6 km) wide at its most narrow part, the Guardian Angel has an area of 361 square miles (936 km^2).[1] Only Isla Tiburón is larger, also situated in the so-called Midriff Region of the upper gulf. Ulloa named neither of these great islands, nor did he set foot on them. He was, however, the first European to reach the delta of the Colorado River by sea and the first to understand that the body of water separating Tiburón and Angel de la Guarda is part of a seaway closed off to the north.[2] Ulloa was not the explorer who bestowed the designation *Mar de Cortés* on the body of water now called the Gulf of California, but there must be some prospective value in naming a sizable geographic feature for one's patron. After a subsequent voyage along the Pacific coast of the Baja California peninsula, Ulloa vanished from history, never to realize any practical gain from his discoveries.

During the early history of the Californias (upper and lower: *alta* and *baja*), treasure was reckoned in the number of native souls rescued by the Church. It was in 1746 that the Jesuit priest Fernando Consag gave the name Bahía de los Angeles to the huge bay that offered a sheltered landing place for provisioning the mission founded 16 years later in 1762 at San Francisco de Borja,[3] located another 22 miles (35 km) farther inland. The large island, reckoned by Francisco de Ulloa to be 12 leagues in length fronting the bay, thus came to be called the Guardian Angel. I first gazed across the strait (Channel of Whales) to the island's massive bulwark during one of my earliest excursions to the Baja California peninsula in 1990. Feeling adventuresome, our group elected to spend two days on Smith Island, located deep within the bay. Because we failed to put our food stocks under proper cover, our supply of fresh tortillas was raided by hungry ravens soon after our arrival. Thus, my initial sighting of the Guardian Angel was tinged with misgivings about being unable to sustain even a short stay on a much smaller island far closer to the amenities of civilization. From all vantage points within the bay, the Guardian Angel is so fulsome that it chokes off any sight of the greater Gulf of California, masked behind the island's towering profile.

Before launching a research trip to Angel de la Guarda in 2007, I slowly gathered some measure of the place from others who passed that way. The brief account by John Steinbeck and his biologist friend Ed Ricketts caught my earliest attention. They stopped off at the island's north end at a place called Puerto Refugio on April 2, 1940, during their historic tour of the Gulf of California on the fishing vessel *Western Flyer*. The unexpurgated first edition of their log, advertised as a "leisurely journal of travel and research,"[4] gives a detailed description of intertidal life found along the rocky shores and sand flats at some twenty sampling stations scattered throughout the Gulf of California. Half their collecting localities can be characterized as rocky shores. Original notes kept by Ricketts record the shoreline's physical intensity at Puerto Refugio: "The rock ledges and rocky pools in which much of the collecting was done were fairly exposed to gulf waves from the north and showed all evidence of being pretty surfswept—considering the limited size of the gulf—more so than any places North of Cape Pulmo."[5] He also noted tidal currents and a sizable tidal range of 13 feet (nearly 4 m) at Puerto Refugio.

A signature concept promulgated by Ricketts is the vertical zonation of intertidal marine life regulated by resistance to wave shock, particularly on outer rocky shores. The idea is expressed chiefly through the novel organization that he and his coauthors adopted in *Between Pacific Tides*.[6] Still available in a revised edition, this masterful treatise sorts out marine life by physical geography, as opposed to a mere systematic taxonomy. Puerto Refugio features a weather-beaten shoreline similar to the extreme setting in which organisms such as the common blue mussel, *Mytilus californianus*, thrive on headlands like the Monterey peninsula on the California coast. Among the marine invertebrates recorded by Ricketts in tidal pools from the north end of Angel de la Guarda was "a species of *Mytilus* different than at Monterey."[7] The survey by Steinbeck and Ricketts was the first of its kind in the Gulf of California. Much was new to them and new to science. However, no subsequent confirmation of this genus in the Gulf of California has ever been made.

The mussels noted by Ricketts may have been juvenile *Modiolus capax*, a species similar to *Mytilus californianus* in size, shape, and coloration. The Golden Fleece hunted by Steinbeck and Ricketts was nothing less than a new ecological philosophy applied to unlocking the web of relationships among marine invertebrates and the physical environments they occupy. *Modiolus capax* is but one species in the fabric of that intricate web as it pertains locally to the shores of Puerto Refugio, regionally to the Gulf of California, and beyond.

Ray Cannon found bigger fish to fry through his angler's adventures in the Sea of Cortez during the 1950s and early 1960s. Author of one of the earliest all-around guidebooks to the Gulf of California and related peninsular shores (Cannon 1966), he not only offered enthusiastic advice to sport fishermen but also gave witness to the region's stark beauty and the powerful intensity of tides, winds, and waves that regularly assault the upper gulf. A frequent visitor to Bahía Refugio on Guardian Angel, Cannon described his first encounter with the place as a cruise "into the outer perimeter of

Dante's Inferno" on account of the bay's dark and jagged rocks, largely barren of vegetation.[8] Fishing in hell, however, was apparently an angler's thrill and an ichthyologist's paradise. In a more scientific frame of mind, Cannon relates how a survey of a small 3,750-square-foot area (almost 350 m^2) of Bahía Refugio that he participated in yielded an astonishing tally of more than 100 different fish species.

A springtime phenomenon that Cannon must have observed within the confines of the Refugio on more than one occasion was what he called a fish pileup, more commonly known as a fish boil. In a boil, the water seethes and erupts with the violent interaction of predatory fish such as the larger yellowtail feeding on schools of sardine and herring drawn to the shallows for spawning. Attacked from the sides and below, the smaller fish typically surge upward and break the surface in a disorderly commotion. Cannon described the flight tactics of the smaller fish as constituting a massive "silver blanket" capable of rising two to three feet (60–90 cm) out of the water.[9] The activity attracts other predators to the roiling water, including sharks, dolphins, sea lions, and pelicans, where they may feed on both the herding yellowtails and smaller fish.

With a dash of hyperbole, Cannon (1966) pokes fun at a pair of midwesterners whose only fishing experience prior to visiting the Guardian's Refugio involved Iowa's largest body of water, Lake Okaboji. This tale is amusing to me, because I grew up in Iowa and I know that particular lake quite well. The Iowans were far from home and clearly out of their accustomed element. Attempting to land a 50-pound (nearly 22.5-kg) yellowtail, the Iowans were startled by a sea lion that chomped on the fish, followed in short order by a bottlenose dolphin in hot pursuit of another great fish that attempted escape by diving beneath the bow of their moving skiff. In Cannon's telling of the story, the speeding dolphin elected to vault over the boat, causing the fishermen to duck down in reaction. Meanwhile, a pelican flying out ahead tucked into a dive after a wounded sardine but ended up crashing into the boat at the feet of the now-terrified Iowans as the boat overrode the pelican's intended target.

I know what it means to be a greenhorn and only gradually become at ease with the varied and sometimes unexpected rhythms of a new place. While I have never witnessed a dolphin jump a small boat, I was once the target of two playful dolphins that overtook my kayak in stealth on a calm morning and suddenly leaped half clear of the water on opposite sides within arm's reach. It felt as if my pounding heart would erupt through my throat, but I was almost immediately overcome by a tingling sensation that spread throughout my entire body in celebration of the sheer joy at having experienced wild nature at close range. I cannot believe that the Iowans of Cannon's tale made their first and last visit to the Gulf of California after their encounters on Bahía Refugio. Fishing on Lake Okaboji could never have been the same again.

Solitude, self-reflection, and renewal may well constitute the most rewarding treasures on reaching a desert island sufficiently large and hazardous enough to dwarf one's ego. The most evocative writing I have found on Isla Angel de la Guarda was composed by Doug Peacock as the prologue to an exquisite picture book on Baja California.[10] A seasoned outdoorsman and naturalist, Peacock relates his experiences during February 1977 on a 10-day sabbatical to the Guardian Angel undertaken with rudimentary supplies that included a sleeping bag, five gallons of fresh water, and little else in the way of food stocks. His intention was to live off the land. Peacock was delivered to the island by the intrepid bush pilot Ike Russell, whose exploits around the upper Gulf of California are legendary.[11] To be dropped into a wilderness place by light aircraft and just as suddenly left to your own devices must rank as one of the most trusting and psychologically challenging things any sane person is likely to do. Russell mentioned that he had spotted boojum trees on some of the island's higher peaks around 4,000 feet (1,219 m). Peacock thought he might get a closer look. Otherwise, he had no particular agenda.

Foraging for an adequate supply of food, in fact, was a time-consuming job. At first, Peacock lived on *Modiolus* mussels easily scavenged from rocks in the upper tidal zone. Although he did not reach Bahía Refugio, he spent most of his 10 days in the far northern part of Angel de la Guarda overlapping the same coastal territory that was familiar to Steinbeck, Ricketts, and Cannon. By carefully observing and marking variations in the irregular semidiurnal tides, Peacock eventually learned to take advantage of sand flats exposed in the lower tidal zone where he could dig for the larger and meatier Venus clams (*Dosinia dunkeri*). By the end of his stay, he had mastered the art of rigging a hand line with a float for successful fishing.

The solitude was immediate, but Peacock found that it took the first four days of his stay before he was able to achieve the inner balance needed to move beyond his personal psychoses and begin to live with the rhythms and realities of the land on its own terms. Although he penetrated inland and attempted to climb high enough to view the boojum trees, the rugged terrain proved too difficult.

Rarely impulsive on my own account, I took advantage of a January 2006 trip with an itinerary that included a stop in Bahía de los Angeles to ask around town about arrangements for future boat support on Isla Angel de la Guarda. If reasonable, our intention was to make the necessary agreements and lay plans for a trip to the island a year in advance. While in town, we learned that Graham MacKintosh was intending to visit the Guardian Angel, perhaps in an attempt to complete a 133-mile (214-km) walk around the island's perimeter. He had provisions and equipment in Bahía de los Angeles ready to go, but strong winter winds made it impossible to schedule a safe trip to the island with a guaranteed extraction date. He had been waiting for a week to make the crossing, with no guarantee about when he might be brought off the island if landfall were possible. We were advised that our own trip would stand a better chance of adhering to a strict schedule if it could be postponed to spring, when the seasonal winds died down. Still dicey in terms of timing, a two-week window of opportunity seemed possible for a trip during a break in our teaching schedules in late March 2007, more than a year away.

Through the following year, I was fixated on MacKintosh and his plans for Isla Angel de la Guarda. Had he managed to reach the island? Did he think he could make the entire circumnavigation on foot at one go, or did he intend to make multiple trips to the island over a more extended time span and hike the coast in segments? I was certain that MacKintosh expected to reinvent Peacock's experience but on a grander scale. Frankly, it was not the feat of physical endurance that stirred feelings of jealousy within me, but rather the prospects for an all-inclusive, firsthand reconnaissance of the island's rock exposures in sea cliffs and adjoining canyon lands. MacKintosh is famous for his 3,000-mile (4,800-km) walk around the entire coastline of the Baja California peninsula, which he described in thrilling detail in his first adventure book (MacKintosh 1988).[12] Knowing it was impossible for our research party to cover the territory the same way MacKintosh might tackle the island, I became obsessed with achieving an equally thorough overview of the place with an eye toward picking the most promising spots for our subsequent geological studies.

How does a naturalist prepare to visit an uninhabited desert island, especially one set in an unpredictable and temperamental seascape? As a matter of public record, ecologist Gary Polis and postgraduate researcher Michael Rose from the University of California–Davis (along with three members of a research delegation from Japan) lost their lives in 2000 during a dangerous crossing from islands in the bay.[13] Tales are legion about boaters caught unexpectedly in the region's rough waters. I am not easily persuaded to make a risky crossing to a mysterious island on the whims of random curiosity. Before I set out for Treasure Island, I expect safe passage and I want a firm grip on the treasure map.

What roles do speculation and serendipity play in the life of a naturalist? They are ever present, but the business of field research makes no assumptions that something unexpected and worthy of interest automatically will turn up of its own accord. Prior to my setting foot on Isla Angel de la Guarda on March 17, 2007, I had recourse to three levels of reconnaissance data. My core impetus to explore part of the island came from plate 1C in the 1971 geological map for the state of Baja California (northern half of the peninsula) by Gordon Gastil and his associates.[14] That portion of the geological map devoted to the Guardian Angel (map 2A) covers less than one-eighth of the map sheet, but the island is rendered at a scale that would fill a standard, letter-sized piece of paper. At such a scale, the map is able to separate the basic components of the island's geology into six colors. The dominant color is brick orange, covering 75 percent of the island. For the most part, the color stands for andesite of Miocene age that girds the island's upland core and steep western coast. Perhaps another 10 percent of the island is painted in pale yellow spreading along the eastern shore, with forked extensions that penetrate the brick orange as canyons representing Quaternary alluvial deposits and outwash on the narrow coastal plain. Little else is left to stir the imagination, except for snippets of pink, blue, dark brown, and mustard-tan colors that fill out the remaining 15 percent of the island.

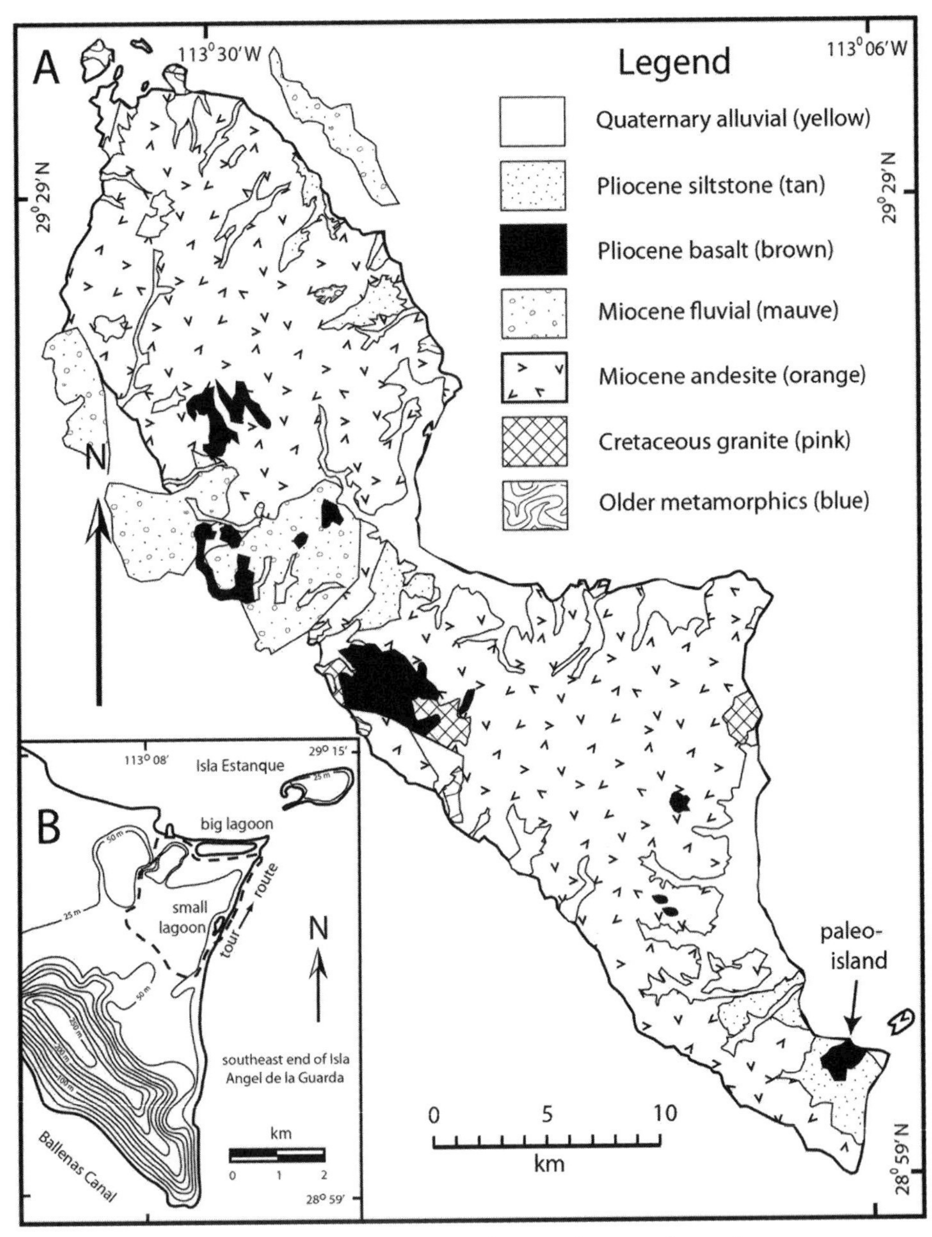

Map 2. Geology of Isla Angel de la Guarda. Map by author.

For me, the key to the treasure map was the brown for basalt surrounded by a mustard-tan color for marine Pliocene strata (map 2A). Gastil and his associates had mapped a paleoisland: an older, much smaller island enthroned on the modern-day island. Some three million years ago, the margins of the Guardian Angel were submerged when sea level stood at a higher level than today. With its own distinct shoreline, the small basalt knob was isolated from the rest of the terrain on the island. I had to see this with my own eyes, because I wanted to know what kinds of marine fossils might be found around the knob's periphery. The quest was now put into play. It remained only to entice my coworkers with the promise that something special awaited us on Isla Angel de la Guarda. The proposition was not hard to sell, because long-time companions on my Baja California travels, Jorge Ledesma and Dave Backus, had participated in similar explorations of paleoislands on the Punta Chivato promontory of Baja California Sur.[15]

The second reconnaissance source consisted of aerial photos shot from a commercial jetliner as it passed over Isla Angel de la Guarda at an altitude of about 30,000 feet (9,144 m) during our return flight to Los Angeles on January 24, 2007. We had been doing fieldwork that month in the Loreto area farther southeast. We were lucky: the sky was clear, and the early afternoon flight was under a high sun with slightly oblique illumination from the west. The aircraft was relatively new, with windows that were clean and still unmarred by the multitude of tiny scratches that gradually dull the outer surface. A better result could not have been achieved had we sat in the cockpit and guided the pilots straight to the target area at the south end of the island.

One of our photos captured the southeast extremity of the Guardian Angel (see plate 2) and included the adjacent islet named Estanque. The bottom-center of the photo picked up the basalt paleoisland mapped by Gastil's team, and it helped convince us that some of the surrounding Pliocene marine strata lapped onto preexisting topography. This was reassuring, because an alternative explanation might have entailed intrusion of younger basalt through older Pliocene strata. The shot also was artistic in the way it framed the juxtaposition of outer and inner lagoons on Isla Estanque and the two closed lagoons on the Guardian Angel. I recall that when we shared this photo with Jorge (not with us on the flight), he advised that we would need to choose a camping place well away from the closed lagoons, so as to avoid any hardship from mosquitoes. This seemed to be practical advice. I was especially gratified that the photos revealed to us the location of the lagoons, because the 1971 geological map gave no hint that anything of the sort was to be found on the Guardian Angel.

Our third reconnaissance source came from a highly technical operation designed to meld a false-color satellite image of Isla Angel de la Guarda with an elevation model over which we could simulate our flight path in three-dimensional space. To implement the project, I engaged a team of tech-savvy Williams College students, who were provided with an appropriate satellite image and asked to solve the problem. Two of those students, Charles (Huajie) Cao and Patty Liao, were with us in Loreto in 2007 as field assistants, and they were aboard the flight to Los Angeles when we took aerial photos of the island. Instead of creating a snapshot or two that provided a bird's-eye view of our potential target area, the team engineered an entire film that followed a flight path from the south end of the island to the north and back again. Viewing the film clip is akin to riding in the bubble of a helicopter that swoops down over a psychedelic terrain shaded in a multitude of colors reflected by the island's variable rock formations and vegetation patterns at wavelengths of light outside human perception. "False color" refers to the assignment of certain color markers to particular bands in the light spectrum beyond our consciousness. For example, areas of infrared radiation picked up by satellite sensors as the reflections of plant life appear not as shades of green but as tones of red. Because the Guardian Angel is a desert island with scant vegetation, this input was practically negligible.

Instead of a landscape carpeted by plant life, the flat snippets of pink (for granite) and blue (for different metamorphic rocks) from the 1973 geological map burst through in the false-color image as a glittering array of tones reflecting subtle variations in rock-forming minerals. The elevation model, which is based on radar technology, imparted realistic three-dimensional depth to the mission. Taking advantage of the lower topography that dissects the rectangular island to form a pass across its midsection, I asked Charles and Patty to plot a figure-eight flight path. Our simulated helicopter trip takes us up the abrupt west coast of the island, crossing midway to the north through the pass to follow the more gentle coastal plain on the opposite east flank of the island the rest of the way to Bahía Refugio. There, our "aircraft" pivots to follow the steep western coast halfway south but turns through the pass to emerge, again, on the east coast. We follow the east shore with its corrugated coastal plain dissected by innumerable dry streambeds toward the south end of the island. We fly over not one paleoisland but a set of small hills separated by what appear to be former marine channels.

Our maiden flight, lasting six minutes, was taken at a leisurely pace, for we did not wish to miss any connection between the island's topography and its geology. We glided so close to the terrain that it was easy to become disoriented and lose all sense of geographic perspective. Thus, we decided to have a small inset map added to a corner of the screen, so our progress through the figure-eight loop could be traced by a red dot that races around its course along the island's margin and through the central pass. I began to associate the red dot with Graham MacKintosh and his potentially arduous pilgrimage to circumnavigate the island on foot. How long would it take him to complete the 133-mile (214-km) course? The little red dot on the computer screen raced the course at a speed a bit better than one-third of a mile per second (536 m/sec).

The unvarnished truth is that no level of reconnaissance work can prepare a person for the reality of landing on a secluded island for a prearranged stay of days or weeks. The boat was loaded with water containers, all our food, our tents, and other camping gear. We had used our aerial photos and satellite image to select what we believed to be the most appropriate camping site. The boatman cut the outboard motor and coasted onto the beach below our predetermined landing spot. In a matter of minutes, the boat was emptied, and the boatman made ready to depart. By the time we moved our cargo off the beach to the nearby mouth of a valley with good access to the island's interior, the retreating boat was only a speck on the horizon. At that moment, the Guardian Angel struck me as the most forbidding and barren of the dozen Gulf islands I have experienced during my career.

Not wanting to make camp after dark, team members separated to erect their tents, then joined together to establish a cooking area and a sheltered depot for the large plastic bottles holding our vital supply of water. The path between the sleeping area and the cooking commons crossed loose gravel strewn with a few mummified carcasses of the spiny chuckwalla (*Sauromalus hispidus*). Endemic to Isla Angel de la Guarda and some neighboring islands, the animal is one of the larger lizards known anywhere in the gulf region. Adult males on the Guardian Angel are larger than the females, reaching a snout-to-vent length (minus tail) of almost one foot (30 cm) and a maximum weight of three pounds (1,400 g).[16] I crossed paths with a particular carcass at dawn every morning before breakfast, and "Chucky Chuckwalla" was the last thing I spied in my headlamp as I set off for bed each night. *Sauro* means "lizard," and the trivial name *hispidus* loosely translates as "hairy, bristly, or spiny." Technically speaking, *Sauromalus hispidus* is the "flat, spiny lizard," not the "bad, spiny lizard." Nonetheless, Chucky always appeared malevolent and never failed to startle me. Perhaps it was the large feet equipped with sturdy claws that made a strong impression. Those strong digging feet allow the chuckwalla to excavate deep burrows, where they escape the heat of the day.

In fact, the chuckwalla is herbivorous and perfectly harmless to humans. Their diet consists mainly of fruits and flowers from one of the several barrel cacti (*Ferocactus peninsulae*) and the silver cholla (*Opuntia echinocarpa*), less commonly the leaves of the common milkweed plant (Smits 1985). The scattered islands in the Sea of Cortez are referred to by biologists as the poor man's Galapagos, on account of the fascinating ecological relationships available for study in a region not nearly as remote and costly to reach. To be sure, one person's flotsam is another person's priceless treasure. The nuances traced by the intertwined life cycles of plants and animals with subtle biogeographic variations across the gulf islands qualify as an ecological Golden Fleece, a magnet for field ecologists specializing in island habitats.

Our visit to the Guardian Angel that March followed one of the island's episodic intervals of extreme aridity, which had reduced the chuckwalla's food supply and led to a population crash. In the absence of larger mammals such as coyotes, ravens and red hawks are the key predators that feed on chuckwallas living on the Guardian Angel, but normally they manage to take only the smaller lizards. The endemic species of chuckwallas are significantly larger than their mainland cousins, possibly because of the lack of big predators on these islands. When times are good and food is abundant, the lizard population swells, and growth increases at a rate such that immature chuckwallas do not remain small for long (Case 2002). Under the stress of drought, however, the die-off is severe, and the birds change tactics to make a living by scavenging on the largest of the dead and dying chuckwallas.

With the campsite secure and much of our first afternoon on the island still before us, the group was eager to move upland through the valley to a place where we could scale the inner valley wall and reach a plateau some 115 feet (35 m) above sea level. There, we were certain to gain perspective over the terrain lying between our target area and us. The paleoislands should be about 1.25 miles (2 km) away and easy to pick out in the distance. More important, it would be wise to reconnoiter for the best route between the campsite and our study area. Climbing the steep valley wall was hard going, but the panoramic view from the top was reward enough for the effort. When we looked back on the path taken, our tents below were the only sign of any human presence. The first of several rocky islets leading to the San Lorenzo Archipelago rose up some 10 miles (16 km) to the southeast, looking like a string of dull pearls cast across a deep blue sea. To the northwest, the paleoislands were discernible beyond a strangely wrinkled landscape of broken rock distinctly orange in tone and reminiscent of the color used by Gordon Gastil and his coworkers to map the island's predominant andesite rocks.

During the remaining hours before sunset, a forced march to the northwest and back again to the campsite was not going to be possible. There were other interesting things to observe in the vicinity. In no time at all, our party dispersed to follow individual pursuits. Scattered about at slightly different elevations amid the rough debris of andesitic rocks were rectilinear clearings covered by loose, silty sediment. My student Peter Tierney went to the center of one such pan-like clearing. He took out his geological hammer and, using the pick end, made an exploratory hole. The sediment cover was deep, and although Peter dug with enthusiasm, no solid rock pavement below the pan could be detected. Surrounding the pan on all sides was a low rampart of well-rounded andesite cobbles and small boulders. Some few boulders were crusted over by a thin, scab-like coating of calcareous material bleached white. What physical processes were responsible for these peculiar configurations?

Figure 4. South end of the little lagoon, Isla Angel de la Guarda. Photo by author.

The others scattered inland to explore more of the strange dirt pans at higher elevations. I retreated toward the valley rim and followed along to a point where it intersected with coastal cliffs. My course led me to an overlook with a deep notch covered in coarse talus descending to the south end of the smaller lagoon we had photographed during our commercial flight less than two months before. The lagoon was isolated from the open sea by a massive berm composed of cobbles and boulders (figure 4). The spot did not appear to be a marshy, mosquito-infested trap, as feared. It looked more like a tidy, artificial lake impounded by the US Army Corps of Engineers. I decided to follow the ravine down to the lagoon for a closer look. The way was steeper than I first judged, making for an awkward descent around and over some of the larger blocks of coarsely broken andesite. What I found on reaching the edge of the lagoon came as a surprise. It was nothing I had anticipated, but I recognized what I was seeing as if it had sprung to life from the pages of a textbook. I was coming into a far country, transported hundreds of million years back in geologic time.

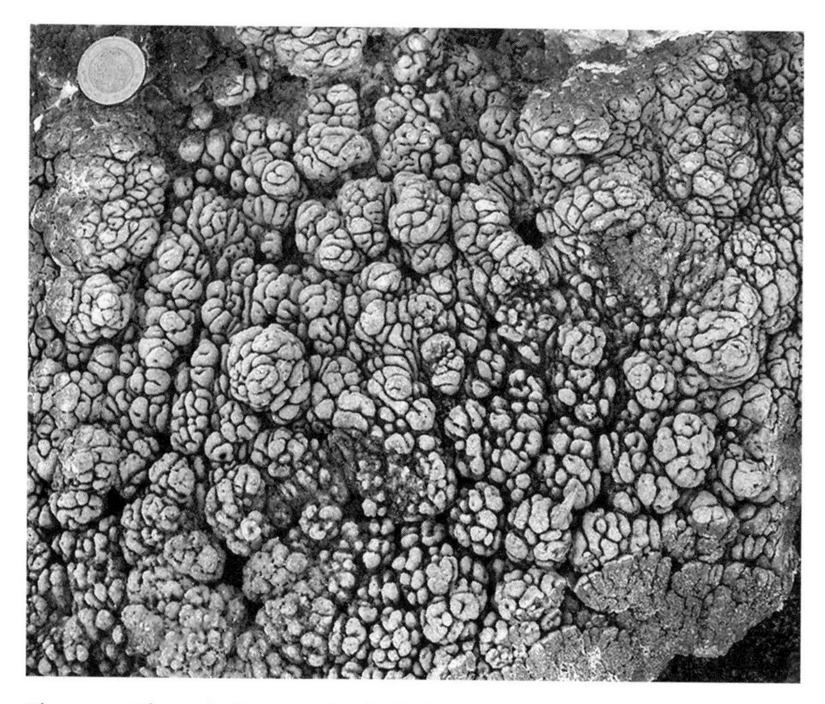

Figure 5. Thrombolites on the little lagoon, Isla Angel de la Guarda. The coin is shown for scale (1 in. or 2.4 cm diameter). Photo by author.

From a distance, it looked like a white rim skirting the edge of the lagoon. One could almost imagine a lake during early winter with a snow-dusted shelf of ice projecting from the shore. But this was no winter setting. Up close, the "ice" rim materialized as a gnarly deposit of solid calcareous material, bleached white under the sun (figure 5). The uneven surface vaguely resembled a mass of upright cauliflower heads, stripped of all green leaves and compressed into a compact sheet. The exposed knobby surface had a raised maximum relief better than one inch (3 cm). The edge of a broken block revealed no trace of internal organization within, such as laminar growth. Here before me was a thrombolite, a structure precipitated by recently living microbes with a long geological history. There was yet more to find beneath the thrombolitic crust. Thin sheets of dark organic material with a distinctly slimy feel formed a soft pad under the crust, perhaps as much as three inches (8 cm) thick. Now, every fiber of my being was jolted awake. These were living stromatolites formed by cyanobacteria with a pedigree tracing back billions of years to a primordial Earth. A quick inspection of the lagoon's shore revealed that the stromatolites could be traced laterally over a distance of nearly 1,000 feet (300 m) along a wide zone up to 6.5 feet (2 m) wide. Stromatolites became known to the scientific world from their extensive fossil record long before sensational discoveries were made of their living counterparts. The earliest technical descriptions of loaf-shaped and club-shaped deposits formed by living microbes were reported from Hamlin Pool in Shark Bay, Western Australia (Logan 1961).

Every college student taking a course in historical geology learns about stromatolites, or the sedimentary-organic structures that form as "stony blankets" having a dome-shaped or often planar architecture. They signify the oldest known vestiges of life on Earth. Indeed, I first learned about them in 1968 in my freshman geology course at the University of Iowa, taught by Professor Brian Glenister. My notebook from that period in my student life records what I was told: "Stromatolites do not secrete lime (calcium carbonate), but instead trap sediment with their sticky cells." Professor Glenister was formerly from the University of Western Australia, where only seven years earlier he had supervised Brian W. Logan for his PhD studies on the modern stromatolites from Shark Bay. I do not recall whether my professor shared any photos from Shark Bay, but almost every textbook on historical geology features an image taken at low tide of the famous domed stromatolites from Hamlin Pool. I suppose I developed a soft spot for this elegant story about modern counterparts to ancient microbial life that thrived in protected hypersaline pools on Shark Bay but seemed to have vanished just about everywhere else. In 1986, I found an opportunity to visit Shark Bay to swim among living stromatolites that formed those classic domes, as well as the simple mats. From that time forward, I proudly shared my own photos of Shark Bay stromatolites with students taking my beginning geology course.

That is how I came to relish a serendipitous moment of discovery on the shores of a small lagoon attached to the far southern tip of the Guardian Angel. If not the first person to enjoy this scene, I was at least among the very first to appreciate its significance. The sun was sinking ever lower on the horizon. The others were sure to have returned to base camp. Now I hastened home along the seashore to find them and share my excitement.

FEATURE EVENT

As Alice Through the "Chukwalla" Hole to Another World

Although exhibiting some disbelief at my news, my companions agreed nonetheless to deviate from our planned objective to march on the paleoislands up and over the escarpment to the north in order to pay a quick visit to the little lagoon immediately after breakfast the next day. That is how we came to follow a more circuitous route to the paleoislands by tracking the coastline around the southeast corner of the island (map 2B [inset]). It is not accurate to say that my claim to the relevance of newfound life on the islands was dismissed without credit, because all of us were familiar with studies on other living stromatolites that had been discovered on the opposite side of the Baja California peninsula along the Pacific Ocean (Horodyski and Vonder Haar 1975; Des Marais 2003). Some of us had actually stopped by the tidal flats at Guerrero Negro to see one of the localities. After the grand scene I had witnessed in Australia at Hamlin Pool, however, I found Guerrero Negro to be a disappointment. The stromatolitic mats existed there, all right, but they were more like threadbare carpets barely buoyant in so much mud. The little closed lagoon near our campsite was something else, something more robust and flagrant. Considering that the matted microbialites on the peninsular west coast made such a stir in the scientific community, it struck me that the first sighting of thrombolites and extensive stromatolitic mats within the Gulf of California would break new and exciting ground.

My companions quickly warmed to the topic, once they saw the place for themselves. The following narrative traces the sights and geological features found along a circuit of about five and a half miles (9 km) around the southeast tip of Isla Angel de la Guarda and back again through the interior to the little valley where our campsite was situated. It is an accurate reflection of what can be encountered and understood on this small part of the island. The details provided, however, required several days of exploration back and forth on various transects and more than one visit to the island to collect additional technical information.

Let us start with the rudimentary facts about the little lagoon. It has a north-south axis that extends for 1,230 feet (375 m), while the east-west width varies from 165 to 250 feet (50–75 m) across, giving the body of water a crudely triangular shape (color section, aerial photo 2) covering five acres (2.25 ha). The lagoon is fully isolated from the open sea by a sturdy berm composed of boulder-size rocks that rise more than 12 feet (3.5 m) above mean sea level. The exposed seaward berm rests on a base about 175 feet (53 m) wide. A less imposing berm is formed by gravel-size rocks on the landward side, also flanked by high cliffs to the north and west. During our visits, the water level within the small lagoon was somewhat below mean sea level. Also, the water is hypersaline: 148 parts per thousand (ppt) as measured along the landward shore. Normal seawater generally has a salinity of 35 ppt. Prolonged local aridity can be expected to promote the lagoon's hypersalinity, while also causing the water level to dip below sea level. The outer berm is very strong but constructed of porous materials that should allow some seepage of "normal" seawater into the lagoon. Major storms could bring some fresh rainwater into the lagoon, especially as outwash from the ravines dissecting the surrounding cliffs at the northwest end. The presence of flourishing microbes in the form of thrombolites and stromatolitic mats around nearly the entire perimeter of the lagoon would imply some degree of relative stability to the ecosystem.

Key to the survival of these microbes is the host lagoon's hypersalinity, which prevents potential invertebrate grazers from taking advantage of a local food resource. That is to say, none of us ever encountered stromatolitic mats in the open lagoons that exist throughout the Gulf of California, where diverse life luxuriates in normal seawater. By comparison, the peculiar hydrodynamics of hypersaline pools located in the inner part of Shark Bay, Western Australia, do permit access from the sea but still prevent herbivorous gastropods from invasion of those pools. Here, as in the world-famous Hamlin Pool, colonies of cyanobacteria maintain extensive structures in a highly restricted ecosystem that allows primordial life to flourish as it had, unmolested, from about 3.8 billion to 0.6 billion years ago before the explosion of metazoan life.

In the kingdom Monera, cyanobacteria are divided into two main taxonomic groups based on cell shape and reproduction.[17] Coccoid cells are globular in shape and solitary. Like all other prokaryotes, they reproduce asexually by binary fission. In contrast, filamentous cyanobacteria have multiple cells arranged in long strands. Cell division remains binary, but walls are erected within the strands to compartmentalize new cells as they are added. When the strands become too long, they typically break apart to make separate colonies. In some species, a long sheath coated in mucus protects the filaments and binds any fine sediment that might be suspended in the water column.

Research based on cultures grown from samples of the dark organic matter collected at the little lagoon was conducted by Maria González in the School of Biology at the Universidad Autónoma de Baja California in Ensenada. Elements of both classes of cyanobacteria were identified by her and reported in our study on the physical geography and biology of the lagoons on Isla Angel de la Guarda (Johnson, Ledesma-Vázquez et al. 2012). The genus *Chroococcidiopsis* was the dominant taxon found in all samples. Solitary cells typically measured between 2 and 4.7 μm in diameter (one μm equals 39.37×10^{-6} inch). Members of this taxon are able to tolerate the effects of desiccation and exposure to high levels of ultraviolet radiation that are normally lethal to other microbes. In nature, these cyanobacteria thrive in arid deserts under extreme variations between high and low temperatures. High brine solutions also are well tolerated by this group. Several other genera belonging to the class of filamentous cyanobacteria were found in samples from the little lagoon. *Phormidium*, for example, has sheathed filaments 10 to 15 μm in length with cell widths of about 1.5 μm.

At low tide and during the right time of day, it is possible to hike in comfort along the gravel beaches and wave-cut platforms below high ramparts of andesite that fortify the Guardian Angel's southeast shore. We are in luck: the tide is out and the morning sun still casts long shadows. The distance from the north end of the little lagoon to the extreme southeast tip of the island is barely one and a quarter miles (2 km). Passing beyond the tidy construction of the outer berm at the little lagoon, one marvels at how nature could shape such a structure. The boulders in the berm are all well rounded by the waves and must have arrived there from the north carried by long-shore currents during storms. The berm's volume, however, is substantial: conservatively estimated as more than 665,000 cubic feet (19,000 m^3). Is there sufficient raw material available at the southeast end of the island to make and maintain such a feature?

The answer is apparent in the geomorphology of the rocky shore traversed by the morning's excursion (map 2B [inset]). Multiple examples of exquisitely sculpted sea stacks are aligned along the shore (figure 6). The outer stacks are squat and stubby, but the inner stacks closer to the cliffs rise to a height of 40 feet (12 m) like fanciful pinnacles, stout at the base, more slender at the waist, and somewhat bulbous at the top. These are sentinels on temporary guard duty, waiting for their watch to end. They fairly scream out the facts of their birth through the powers of wind, waves, and tides steadily eating away at the coast. One by one, they will eventually tumble, reduced by the waves into greater and lesser bits that become worn in the

Figure 6. Sea stacks on the southeast shores of Isla Angel de la Guarda. Photo by author.

surf. Conveniently, the tilted andesite rocks exposed behind them in the cliffs provide the right spacing between layers from which wind-driven waves persistently press and erode to carve out new sea stacks.

All andesite rock begins in volcanic flows. Sometime after extrusion, these flows were set on end by tectonic forces that shaped the Gulf of California. Furthermore, these particular flows are naturally blocky in character, known by volcanologists as agglomerates for the predominance of angular or slightly rounded lumps they contain. Pockets filled with finer, less-resistant material often are lodged among the greater lumps within the flows, providing weaknesses where accelerated disintegration will occur. If the zone of erosion along which the sea stacks are most actively being formed is only three-quarters of a mile (1.2 km) long but 50 feet (15 m) deep, the volume of rock represented by the removal of all sea stacks 40 feet (12 m) high would amount to 8 million cubic feet (more than 225,000 m^3). It means that more than sufficient material has been available through the last tens of thousands of years to remove from the shore and add to the impressive outer berm at the little lagoon.

Another feature that begs for our attention is located near the far southeastern corner of Isla Angel de la Guarda. It can be seen to best advantage (see plate 2) as a small cove 165 feet (50 m) wide near the east end of the big lagoon. We must enter this cove to gain access to the big lagoon itself. Especially from the air, the small cove gives an impression that the US Army Corps of Engineers has, again, been quite busy on the island. The area around the cove is nothing less than a functional spillway that allows excess water to empty into the Sea of Cortez from the large lagoon during rare but episodic storms of near-hurricane intensity.

The dimensions of the large lagoon at this end of the island (see plate 2) dwarf the little lagoon near camp. It is almost perfectly rectangular in outline: three-quarters of a mile (1.25 km) long from east to west and two-eighths of a mile (0.25 km) wide from south to north. In surface area, the big lagoon covers 55 acres (22.5 ha). Hence, it is more than ten times the size of the little lagoon. The big lagoon is fully closed off from the sea by an outer berm composed of gravel and cobble-size rocks elevated more than 12 feet (3.5 m) above mean sea level on a base about 165 feet (50 m) wide. Walking out a short distance along the berm's broad shoulder yields an impressive vista, as one sees it arc forward ever so gently toward the low hills in the far distance. To our left, the lagoon water is pale blue, while the water to the right in the deeper open sea is a darker shade of blue. The stature of this massive berm built by nature's hand is awe inspiring. Satellite imagery indicates that there is no fully enclosed lagoon any larger than this in the entire Gulf of California.

We retrace our steps eastward to exit the outer berm and begin to trace the lagoon's inner shores westward. An inner berm encloses the complete south margin of the lagoon, rising to an elevation about 24 feet (7.25 m) above mean sea level. A veneer of pebble-size rocks covers the ridge. Crossing inland over this structure, we drop into a dry moat that extends the full distance parallel to the big lagoon. The floor of this depression is pan-like and devoid of any vegetation. It sits at an elevation about 17-plus feet (5.25 m) above mean sea level. Farther inland are yet more pan-like depressions at increasingly higher elevations. All together, they appear to form an ascending series of raised lagoon bottoms. That is to say, the island was subjected to tectonic uplift that left behind a succession of dry lagoons similar in shape to the big lagoon but not always as large.

Back toward the big lagoon, distinctive scour marks are visible in the loose gravel on the inside face of the great outer berm as we draw nearer the opposite shore. These marks were left by storm waves that washed over the top of the protective berm, probably coincidental with high tide. They are as clear as trackways left on a damp beach that show which direction a bird walked. Viewed directly in front of us, the scour marks are slanted from the top left of the berm to the lower right, showing that the waves did not strike straight on from the north but instead struck at an oblique angle from the northwest.

More surprises await us at the edge of the big lagoon. The same calcified crust encountered at the little lagoon covers a vast expanse from 6.5 to 10 feet (2 to 3 m) wide along much of the big lagoon (figure 7). Under the baking sun, the white crust is incised with desiccation polygons that vary from 16 to 30 inches (40 to 75 cm) in diameter. Outlined in black, the cracks penetrate downward to some depth. Gently peeling back the crust on one of the polygons reveals the presence of dark organic material in matted layers up to three inches (8 cm) thick. The living stromatolites are extensive. Hesitantly, we step closer to the open water in the lagoon. Still some distance from the open water,

Figure 7. Polygonal cracks impressed on stromatolitic mats around the inner margin of the big lagoon, Isla Angel de la Guarda (looking west). Photo by author.

the crust and its hidden stromatolitic mat float off the bottom of the lagoon. It is hard to believe, but the sensation is that of walking out on a bog around the margins of a boreal lake in Maine or Minnesota. The crust bends ominously beneath a person's weight and one feels helpless, as if about to break through thin ice on a frozen lake. Water laps over the crust at the edge, where it is no longer bleached white but appears pink. Here, perhaps, the crust's surface remains alive. At the edge, the lagoon water is turbid green. One cannot see clearly to any depth, but the lagoon bottom drops off rapidly.

Samples have yet to be tested, but *Phormidium*, one of the filamentous cyanobacteria identified from the little lagoon, probably has a significant presence here. Elsewhere in lakes and rivers, *Phormidium* has a tendency to occur in mats of tangled filaments that detach from a firm substrate and float to the surface. In France, neurotoxins have been reported to cause death in dogs that consumed matted river microbes with high concentrations of *Phormidium favosum* (Gugger et al. 2005). As extraordinary as they are, it may be that the stromatolitic mats on the Guardian Angel also are deadly poisonous. This could be another reason for their survival on an already hazardous island. More than anything else, the broad and vigorous development of prokaryotic life on the island makes a profound impression.

Skirting the largest closed lagoon in the Gulf of California takes some effort in real time and requires equal mental exertion to wrap one's mind around a stout central trunk in the ancestral tree of life. In her elegant tribute to the prokaryotes, Lynn Margulis (1984) writes about the bias that nearly all naturalists show against "lower" life on the evolutionary scale, through obsession with eukaryotic plants and animals. She reminds us that prokaryotic microbes prepared the world for "higher" life through metabolic exchanges with the atmosphere that produced free oxygen and forever altered the planet. She gives us further pause to think: "Without the prokaryote's continuing activities, neither we nor the animals and plants on which we directly depend would continue to exist."

Teasing us with the enormity in contrast between "lower" and "higher" life at two adjacent settings, a pristine little bay sits open to the sea less than a half mile (0.75 km) west of the big lagoon. We reach it by staying on the coast until we encounter a small headland thrust into the sea like a fat thumb. After crossing a wide saddle between the headland on our right and the rocky hills rising on our left, we drop down to a corner of the bay through a dune field composed of dazzlingly white sand. The reflection off the sand under the midday sun is so bright it hurts the eyes. The water in the little bay is a soft turquoise blue, more soothing to the eyes. A green turtle is startled by our approach and retreats from the shallows. It is a perfect spot for lunch. While we sit on the beach, letting the clean sand sift through cupped hands comes naturally. These beach and dune sands are almost exclusively calcium carbonate in composition, derived from the reduced shells of countless mollusks, mainly clams and gastropods. Life in the bay is mainly infaunal in habit, as exemplified by the Venus clams (*D. dunkeri*) harvested by Doug Peacock during his stay on the island. At low tide, a multitude of small holes mark where the siphons of the clams reach the surface. It is quiet now, but empty shells regularly are washed onto the beach by wind-driven waves that grind the shells into sand. Satellite images pinpoint 52 places on the Baja California peninsula where more substantial beach-dune complexes occur (Backus and Johnson 2009). As here, dune ripples and other features confirm that the prevailing wind comes out of the north.

We are geographically midway through our circuit around the principal attractions in the southern part of the Guardian Angel. The paleoislands that brought us here in the first place are left to explore. Pointed south with our backs to the turquoise-blue bay, we find a chasm between two rocky hills that invites passage to the other side (map 2B [inset]). The rounded hills, which rise to elevations more than 165 feet (50 m) above sea level, are the basalt inliers mapped by Gordon Gastil and associates. A dry streambed emerges from the hills, but in the foreground it meanders through stratified banks formed by enormous quantities of seashells. Disarticulated shells belonging to the little bivalve *Chione californiensis* are common on one bank. Another features Fischer's oyster (*Ostrea fischeri*), also disarticulated and bone white. These loose deposits are about an eighth of a mile (400 m) inland and 15 feet (4.5 m) above sea level. They represent a high stand in the Late Pleistocene sea level perhaps 125,000 years ago. Wedged near the mouth of the opening, the shell banks demonstrate that seawater flooded at least partway into the passage.

The floor of the arroyo rises smoothly to the south. Disappointingly, no other fossils are spotted within the passage. Once through the other side about three-fifths of a mile (1 km) from the coast, the streambed becomes diverted by stratified deposits of siltstone that truncate against the flank of the basalt hills. These Pliocene deposits form an embankment 46 feet (14 m) high that curves inland around to the west. The streambed is entrenched only against the cliffs and barely contained by the more open ground to the east. Large Pliocene oysters (*Ostrea californica osunai*; Hertlein 1966) are abundant. Smaller whole valves and broken pieces from valves originally 8 to 10 inches (20–25 cm) in length (figure 8) have washed out from the embankment into the arroyo. Less common are pecten shells and rare sand dollars (*Encope angelensis*). It is gratifying to find these fossils, but the diversity is lower than expected. Some three million years ago, a sheltered muddy bay flooded an area nearly as large as that covered by the big closed lagoon today.

Moving up the arroyo, I spy the flimsy, paper-like skin shed by the large Angel de la Guarda rattlesnake (*Crotalus angelensis*) endemic to the island.[18] Perhaps the animals are still in hibernation or under

Figure 8. Fossil oysters of Pliocene age from the south end of Isla Angel de la Guarda. Photo by author.

severe stress like other reptiles on the island this season. The landscape seems devoid of all life. We have a mile and a quarter (2 km) to traverse over rough ground rising to the southeast to meet the top of the escarpment above our tents. Although the slope is gradual, the long steps from one rough andesite boulder to another are strenuous work. Returning full circle to the area explored during our first afternoon on the island, we take time to confirm the elevation of the highest rectilinear clearings covered by loose, silty sediment. The similarity in form to the abandoned lagoons adjoining the big closed lagoon to the north is striking. This one, however, is 125 feet (38 m) above the nearby little lagoon. Is it possible that this feature was formerly a small closed lagoon that sat near sea level and slowly became uplifted to its present elevation through regional tectonics?

A few days after leaving the Guardian Angel, I find myself sitting in a motel room in Ensenada. It will take another day to cross the border to San Diego and catch a night flight home to Massachusetts. I am more excited than I can remember from any previous trip to the Gulf of California. We went out to the island seeking treasure. We were well informed, but the natural plunder we anticipated was not as fulfilling as the riches we stumbled on through outright serendipity. It dawned on me that earlier visitors to the region may have failed to explore the closed lagoons, because from boat level on the open sea it is not possible to see over the tops of the high berms. I take out a loose-leaf notebook and write by longhand 20 pages of text toward the first draft of a scientific paper about the stromatolite lagoons on Isla Angel de la Guarda.

It proved to take rather longer than I hoped before the cursory biological data became available, but I could not stop myself from beginning a new and inspiring facet of my lucky vocation. More questions remain to be answered: How many other closed lagoons in the Gulf of California have resident microbial mats? If we assume that the lagoons formed at different times over the past hundreds of thousands of years, how did the species forming the mats arrive there? Most intriguing of all, the primitive prokaryotes surviving on Earth raise questions among exobiologists regarding possible life on other planets.

3

Great Sand Ramp at Km 41

No man is less aware of things than the conscientious traveler who hurries from wonder to wonder until nothing less than the opening of the heavens on judgment day would catch the attention of his jaded brain.

—*Joseph Wood Krutch,*
Baja California and the Geography of Hope

ARID LANDSCAPES OCCUPY about 9.5 percent of the planet's land surface, with somewhat more than half that area dominated by dune fields.[1] Desert sand, therefore, buries more than 2.7 million square miles (7 million km^2). This is roughly equivalent to an area more than three times that of Greenland. As a general rule of geography, however, many of the world's great deserts are dispersed on the west sides of continents and centered at latitudes between 25° and 35° north or south of the equator. The Namib, Atacama, and Gibson deserts of Africa, South America, and Australia, respectively, all occur close to 25° latitude in the Southern Hemisphere. Technically, much of Baja California is an extension of the Sonoran Desert. Various desert subregions are defined as separate entities throughout the peninsula (Roberts 1989). Recording an average annual rainfall of only 2 inches (5 cm), the San Felipe Subregion represents the driest part. Here, at around 31° north, vast sandy plains, alkali flats, and sand dunes dominate the lands adjoining the Gulf of California.

The drying is only moderately less severe farther down the coast. What is called the South Gulf Coast Desert Subregion follows below Bahía de los Angeles (Roberts 1989), where the local flora of cardón and cholla cacti, as well as creosote bush and brittlebush (*Encelia farinosa*) together with shrubs in the leatherplant family (Euphorbiaceae) is better established. The gravel road leading south out of Bahía de los Angeles turns southeast to cross the spectacular, mountain-rimmed Valle las Animas oriented north-south, but open to the Gulf of California at Bahía de los Animas. On descending the rough road across this valley, the traveler is confronted with a natural wonder that may as well rival the opening of the heavens (figure 9). A massive sand ramp climbs the south end of the valley and spills over the leading edge of mountains through a low opening slotted in solid ramparts. Skirting around the outer toe of this immense structure, the road maintains a distance not less than five miles (8 km) from the nearest dune heights. The vertical height of the ramp sand plastered against the naked rocks appears to be about 245 feet (75 m). One might at least approach heaven by climbing such a grand incline as this.

Figure 9. Gravel road south of Bahía de los Angeles, showing the great sand ramp in the background. Photo by author.

There exist other, more worldly reasons for wishing to hike across such a commanding topographic feature, although the irreducible fun of the experience cannot be so easily dismissed. From a scientific viewpoint, the coastal sand dunes scattered along the Gulf of California's peninsular coast are special, owing to the fact that sand composition is dominated by calcium carbonate ($CaCO_3$) derived from the minutely crushed shells of mollusks and other marine organisms available in vast quantities. Sand dunes strung along the opposite Pacific shores of the Baja California peninsula may be more continuous and voluminous in size, but they are overwhelmingly composed of inorganic materials dominated by quartz grains (silica: SiO_2) and feldspar (complex potassium and sodium-aluminum silicates). On both sides of the peninsula, the coastal sand dunes are intimately associated with sandy beaches. Deflation of beaches by the prevailing winds operating over these separate quarters is the ongoing process by which sand is carried off the shore and transported inland.

To understand why the sand dunes on the gulf coast are fundamentally different from those on the Pacific coast, it is necessary to first consider the sources of sandy beaches on opposite sides of the peninsula. In a comparative study of 50 such beach localities equally divided between the east and west coasts of the peninsula, Carranza-Edwards and colleagues (1998) were the first to demonstrate the fundamental compositional differences between the two shores. Like their associated sand dunes, the beaches on the gulf side are unusually enriched by organic shell debris, while on the Pacific side the beaches are more purely composed of silica.

The controlling factor over this disparity is the raw physical geography of the peninsula (see map 1). Steep fault escarpments define craggy coastal ranges that dramatically hug the eastern shore, in contrast to the low coastal plains that confine a succession of huge lagoons on the western shore. Although only intermittently activated by running water, the tributaries on the west side of the divide follow longer watercourses and, thereby, erode rocks over a much larger area than do the shorter streams descending to the Gulf of California. In other words, substantially more ground-up rock reaches the broad plains and beaches on the Pacific coast than is dumped onto the gulf coast. With little contamination from inorganic materials, the confined pocket beaches on the gulf coast tend to be overwhelmed by organic shell debris, collected and reduced into smaller and smaller fragments in rough surf stimulated by the winter's north winds. In brief, the sand spits and related dunes on Pacific shores are fed from the land, while many of the isolated beaches on the gulf coast are nurtured from a beneficent life spawned in the Sea of Cortez.

A strenuous hike would be necessary to reach the most remote stretch of sand piled against the high flanks of the mountains inland from Bahía de los Animas (map 3), but the trek promised to answer an intriguing question. Was it possible that carbonate sand derived from the minutely fragmented shells of bivalves and other mollusks concentrated on the beach at Bahía las Animas might be blown as far as 15 miles (24 km) inland by strong winter winds? I had to know the answer, and the only way to get that answer was to climb to the top of the sand ramp. The exigency of such a mission was not immediate but grew by increments in my mind over several years until it became a gnawing irritation.

My first encounter with the sand ramp occurred in January 2002, when our small research party ventured south from Bahía de los Angeles with the goal of reaching the embayments around San Francisquito by nightfall, a distance of only about 80 miles (129 km). An early start from Bahía de los Angeles was delayed by the necessity of restocking for food at the grocery store, by consultations with locals about the status of the road ahead, and by a visit to a mechanic's shop to seek advice on the proper tire pressure for travel on the long gravel road that lay ahead. It was close to midday before we were finally ready to depart with our rented Isuzu Trooper, now fully loaded with supplies. It was the first trip for all of us on this back route to the south, and the driver was cautious. We were in no great hurry. Leaving Bahía de los Angeles, the road follows the narrow Valle las Flores, confined by the Sierra las Animas, formed mostly by tilted andesite blocks rising 3,250 feet (1,000 m) to the east, and the yet higher Sierra San Borja, with its massive granite pluton reaching upward more than 4,600 feet (1,400 m) to the west. Little more than 9 miles (15 km) out of Bahía de los Angeles is the abandoned mining village of Los Flores, where gold and silver ore from deep within the San Borja mountains was milled for transshipment up until 1934. The town cemetery features several generations of the Daggett family, whose patriarch was a ship's officer who hailed from Oxford,

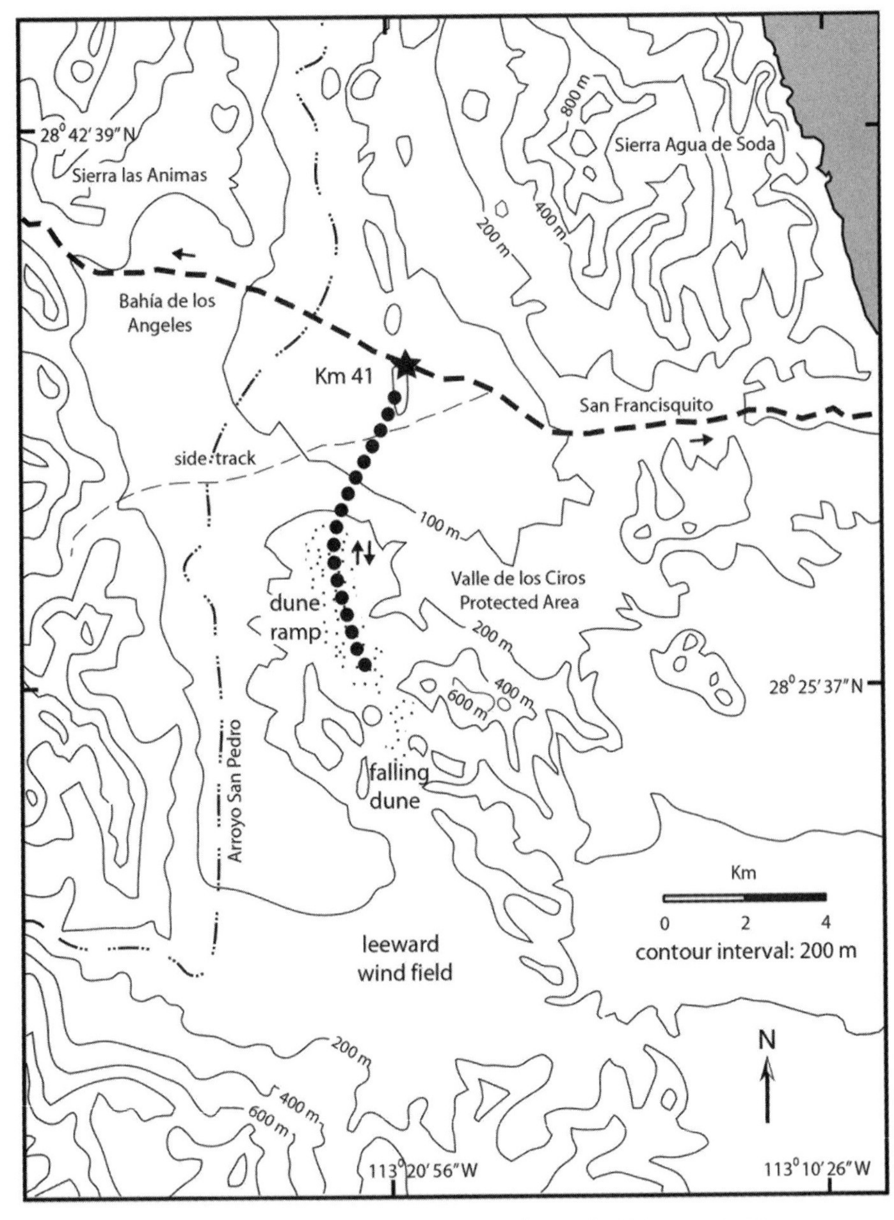

Map 3. Topography of region around El Quelital (Km 41). *Note:* Star marks location of kilometer 41; dashed lines show primary and secondary roads; and the dotted line marks the route of the trek to and from the great sand ramp. Map by author.

England, and jumped ship in the 1880s.[2] We paid our respects at the grave of Dick Daggett Jr. (1893–1969) and moved on.

The first view of the great sand ramp emerges around a bend in the road on exiting the Valle las Flores. Our studies on the carbonate dunes of Punta Chivato (Russell and Johnson 2000) were already two years behind us but still fresh in our experience. Here before us were the largest sand dunes any of us had met along the gulf side of the Baja California peninsula. The attraction was immediate. Stopping the vehicle, we jumped out to snap photos. I recall climbing atop the Isuzu to gain some slight elevation. Progress was slowed by frequent subsequent stops at places along the road with potentially better camera angles. The road dropped in elevation and maintained a steady distance below the dunes. Clarity of the air diminished as the sun sank lower in the west. In the hazy light, the perfect curve of a central dune almost appeared like a wide road rising to a gap in the mountains.

Was there actually a passable connecting road in that direction? We slowed to a crawl and all necks craned southward in search of an easy track on which to depart the gravel road and approach nearer the dunes. At long last, we found such a track, situated far to the west, well beyond the most distinguished-looking dunes. We took a quick vote and decided to turn off to reconnoiter the dirt path. Other vehicles clearly had followed this way before. Ours quickly sank into soft dirt a short distance off the main road. The tires of the Isuzu would need to be deflated quite a lot more, before forward traction might be gained. We grudgingly agreed that retreat was the better option and managed to push the vehicle onto firmer ground. The day had grown late, and we decided to camp on the spot. We were less than halfway to Francisquito and now behind schedule. I laid out my mat and sleeping bag on the track behind the vehicle, where I slept secure in the thought that another vehicle would not arrive in the night to crowd our foolishness.

Another four years passed until January 2006, when I encountered the great sand ramp for the second time. On this occasion, our research group made the decision to return to the embayments around Francisquito to complete a mapping project. Once again, Bahía de los Angeles became the jumping-off point for the excursion. As our field party needed to accommodate three students in addition to the usual cast of characters, we made preparations to have two vehicles at our disposal. Driving in tandem on a gravel road kicks up a great dust plume that forces the driver of the second vehicle to follow behind at a considerable distance. Before setting out from Bahía de los Angeles, however, we agreed that those in the lead vehicle would try to locate a side road to bring us closer to the dune field. Maps showed that Arroyo San Pedro drained the region around the dunes to the west. Our vigilance in searching for a sidetrack had been instigated too late on our previous attempt, because we apparently had missed some promising paths to the interior along this drainage. By making an early morning start from Bahía de los Angeles, we thought we might have some extra time to reach the dunes before continuing the journey to Francisquito.

When a viable sidetrack was identified from the lead vehicle, the hunt was on. Leaving the second vehicle behind, we crowded together and proceeded up the well-used track under the confident engagement of four-wheel drive. After a good many minutes of driving, it became apparent that our path tracked back parallel to the main gravel road but more to the south. We were not getting much closer to the actual dune field. When the track crossed the principal arroyo of the San Pedro, we decided to turn and follow a set of tire tracks upstream. This brought us nearer to the dune field, but we were still a considerable distance off on its west margin. Where the four-wheel drive began to suffer some hesitation, we halted and then carefully turned the vehicle around to point back downstream. After we climbed the embankments of the San Pedro, we made a half-hearted attempt to push ahead on foot in the general direction of the dune field. The vegetation was somewhat congested around the borders of the arroyo, which made marching in a direct line all but impossible. It was now close to midday, and our group arrived at a consensus that the effort to reach the dunes had to be abandoned if we were to make Francisquito in sufficient time to set up camp that evening.

It was only an additional year, until January 2007, before the next opportunity for action presented itself. Our excursion to Isla Angel de la Guarda was the premise for another road trip to Bahía de los Angeles. The truck from the motor pool at the Universidad Autónoma de Baja California that brought us from Ensenada to

Bahía de los Angeles sat unattended, awaiting our return from the island. When we came off the island with a day to spare before our scheduled return north, I coaxed Jorge to take us south on a special mission to conquer the sand ramp. It was a hard argument, because the truck was not meant to leave the paved road.

Instead of checking into a comfortable motel in town after crossing the Ballenas Passage from the big island, we transferred our camping gear and remaining food stocks to the truck and set off directly for the abandoned mining town at Los Flores. There, we pitched tents and prepared an uninspired camp meal. With firm assurances from me that under no circumstances would the now-wayward truck leave the confines of the gravel road, we packed up the next morning and drove south at a conservative crawl. The plan was to find the most convenient spot on the main road from which to launch the final assault on foot against the great sand ramp. That spot was reached at the roadside marker for kilometer 41 (25.5 miles south of Bahía de los Angeles), where a right-angle turn on a well-trod set of wheel ruts brought the vehicle over a low rim and onto a small field carpeted with stubble grass. Here Jorge would remain with the vehicle, while the rest of us set out for the sand ramp.

FEATURE EVENT

Like Monks Marching in Holy Procession

We are fortunate. The day is somewhat overcast with a thin cover of clouds that promises to keep the day mercifully cooler than it might become otherwise. Round trip, this will be a 10-mile (16-km) hike. We fill our water bottles and pack some fruit and sandwiches for lunch. The object is to travel light, so we empty our daypacks of all nonessential items. Cameras and sunscreen are essential. Rock hammers and assorted guidebooks and notebooks are not. Dave and I and our student, Peter, make a determined threesome who contemplate the sand ramp above us on the distant horizon. I look at my watch: the time is exactly 10:15 a.m.

After 20 minutes walking downhill from the roadside, we meet the well-established track that we had driven on only a year earlier. Now, instead of keeping to the road, we cross it and plunge ahead due south. There is a sense of triumphant determination in our steps. The prize is before us, and we will not be denied the outcome we desire. Here, the landscape supports scattered stands of small to medium-size cardóns, as well as the occasional organ-pipe cactus (*Lemaireocereus thurberi*) and a diffuse spread of palo adán (*Fouquieria diguetii*). The latter, also known as Adam's tree,[3] has fearsome, stiff spines that glove its long, willowy branches from end to end. Where possible, we give them a wide berth. Where skirting them is impossible on our line of march, the first through a tangle of palo adán must take great care not to release a bowed branch against the person coming behind. Those following instinctively raise one elbow to cover the face and hold out the other arm to receive the offending branch with fingers held like forceps to grasp a narrow bit of available stem between thorns. This style of hiking requires fine motor skills for precise hand-eye coordination.

Climbing out of the gentle defile in which the track is situated, we join a more open country that rises ever so slightly to the sand hills in front of us. The stately cardóns fall behind us, and the dominant palo adán is largely replaced by sentinels of silver cholla that stand high on erect trunks like small trees. These, too, must be treated with considerable respect. The thorns of the cholla are not as stiff as those that protect the palo adán, but they are hooked and easily grab onto any piece of clothing that brushes too close. Where a thicket of cholla cacti blocks our path, there is no choice but to make a wide detour. Instinctively, we are drawn to more-open land where the creosote bush and the brittlebush occupy territory. Perhaps aided by strong tannins, the creosote bush maintains an even separation from other bushes of its kind and repels crowding by competing plant species. There is no such thing as a tangle of creosote bushes. This makes for more pleasant walking through the countryside, although the line of march quickly shifts to a zigzag pattern. One could run through a field of mature creosote bushes without disturbing a single branch, but doing so in a straight line, or even along a smooth arc, is impossible.

Little more than an hour and a half has expired since we left the vehicle, and it seems as though we have made barely any progress to reach the lower hemline of the skirt around the great sand ramp. We are determined in our goal, however, and the pace quickens where

Figure 10. Vegetation skirting the great sand ramp: Dave Backus and Peter Tierney march through a field with sandburs. Photo by author.

the fixed density of creosote bushes gives way to more open ground thinly blanketed by yellow-green grass. The first impression is that cattle must be somewhere nearby, for the place has the look of well-grazed pastureland. Gaunt-looking cholla trees with upraised limbs held close about them stand widely scattered here and there. But the vegetation cannot obscure the great sand ramp (figure 10), sitting fully in front of us. Still, the ground immediately ahead remains almost flat.

Not having to deviate from a straight course, we begin to make progress, with each long stride appearing to bring us closer to the base of hills from which the great sand ramp rises. Just here, however, another factor enters the equation against which progress in time and distance must be calculated. At first, it is a minor inconvenience, causing the pace to slacken only slightly. The low grass over which we cruise forward with every footfall is loaded with burs mounted on thin stems projecting four to six inches (10 to 15 cm) above the ground. The burs are spherical husks formed around seeds. Hairy prickles that radiate out in all directions cover the husks, giving them an exaggerated bulk three-quarters of an inch (2 cm) in diameter. The burs easily detach from their stems, to be carried away on the feet and legs of unwitting animals. Geologists unschooled in botany also fall prey to the plant's clever trick of seed dispersal. It is possible to cross a good distance before the first burs begin to migrate up the pant legs of thick jeans. The leading bur climbs by first attaching to one cuff or hem of the trousers. When the opposite trouser leg brushes against the first, the bur transfers from one limb to the other. With the lifting motion of each leg performing its part of a forward stride, the bur is naturally transferred a little higher as it moves from one trouser leg to the other. This migration may go undetected until the bur climbs above the knee and advances halfway toward the crotch. When this stage of discomfort is detected, even a highly motivated geologist will stop to find out what the problem is. An added complication is that fingers are no help against the clinging capabilities of burs from this species. The dreadful things stick to fingers and are not easy to be rid of.

The grass in question is probably a species of sandbur (*Cenchrus palmeri*), with the Baja California representative possessing the largest burs found anywhere in the same genus.[4] Humans, of course, are adept at tool making. Where fingers are incapable of fixing a problem, some sort of substitute must be found. But by the time a suitable solution is in hand, so to speak, the inner thighs are likely to be crowded with burs that continue to climb ever higher as long as the owner continues to move forward. What appeared at first to be an expressway to the great sand ramp now has turned into a debilitating hazard course. Humans also are adept copycats who have only to observe a new trick at tool making by its first inventor, before it is spread among the entire tribe.

A stick of wood from the skeleton of a cactus makes the perfect tool with which to bat away burs from any part of the body. Not any stick will suffice. The most suitable tool must be long enough to swing through a sufficient arc so that the distal tip impacts the bur with such force that it cannot adhere to the wood. Through trial and error, flimsy sticks that easily break are rejected, while stronger sticks that stand up to a good swatting action are selected. Not surprisingly, the sturdy but thin ribs of the cardón make the best bur bat. However, cardóns are less abundant here than earlier on the trek, and dead cardóns are downright scarce.

After some delay, all members of our little tribe are outfitted with skeletal cardón ribs, which also make superb walking sticks. The march continues onward, although the tribal procession takes on a decidedly peculiar appearance. Skill in bur deflection requires the practitioner to quickly switch from walking-stick mode to swatting-stick mode once a bur is spotted clinging to the boot or lower trouser leg. The stick must swing down, forcefully swiping the shoe or ankle. It means that one must walk with head downcast, ever alert for the next bur. If not steadfastly guarding against the initial creeping invasion of persistent burs, the practitioner must swipe at the lower trouser legs, an action that makes the inner calves a target for an unmerciful beating. Under no circumstances may any bur be allowed to reach above the knee.

How, then, must such a parade appear to the uninitiated? Were we clad in hooded robes made from sackcloth, I fear we might be taken for a line of penitent monks in some religious ritual of purification engaged in self-flagellation. Heads bowed, we walk briskly forward in silence. The only noise is the rhythmic beat of the sticks against our bodies. Precious minutes slip away; it is already past midday and the expected climb through sand has yet to start. The morning's cloud cover has dissipated to thin wisps that no longer restrain the day's growing heat. It is warm, and our water is consumed at a faster rate, with more-frequent rest pauses. We near the southern edge of the grassy plain to find that the land drops again, before it truly joins the first hillocks of brown basalt with intervening piles of sand. On the downslope, the cholla trees join ranks to form a more closely guarded barrier. Behind them stands a crowd of palo adán that retards our progress. It is 1 p.m. before we find slim shade below the rock face of a basalt outcrop a short distance up the slope. Peter decides to stay put and conserve energy, but Dave and I press onward.

We are barely 750 feet (230 m) above sea level when the serious climbing through loose sand begins. Initially, the slope is gentle. Sparse palo adán and the thinnest cover of grass anchor the sand that fills gaps between basalt spurs. Within a half hour, we are confronted by open sand that forms a steeper incline at the angle of repose, and

Figure 11. Dave Backus seated on a dune near the top of the great sand ramp. Photo by author.

we ascend laboriously to the top of a transverse dune with parallel sand ripples aligned with an opening in the range to the immediate south. Here, at an elevation of about 1,000 feet (305 m) above sea level, we join the wide highway of sand that can be seen so distinctly from the actual road below. The wind gap in the mountains is farther ahead. To get this far has taken us longer than expected; my watch reads 1:30 p.m. Dave scoops a sand sample into a plastic bag and sits down to rest (figure 11). I collect sand in the cupped palm of one hand and bow my head to examine the grains through my hand lens.

As expected, the sand is fine grained. I recognize no telltale traces signifying tiny bits of broken mollusk shells. Indeed, the sand seems to be primarily quartz in origin. A more thorough examination by microscope must be done later. Peter awaits our return, below, so we eat quickly. With Isla Angel de la Guarda rising abruptly on the far horizon, the view north to Bahía de los Animas in the middle distance is breathtaking (see plate 3). Whether or not carbonate materials form any fraction of this sand body, we give mute witness to the simple fact that powerful winds blowing off the Sea of Cortez are responsible for sandblasting the mountains and, indeed, moving a high volume of materials through to a falling dune on the other side. The dry region spread below us from which this material is scoured encompasses a wind corridor with an area easily covering 87 square miles (225 km^2). That nature performs such wonders of deflation and monumental conflation is more than worthy of notice. On this clear day under still air, we find ourselves tired, but exultant, near the summit of a great Earth temple.

My first water bottle is empty, and I have started on the second. It will be difficult to ration the remaining water on the return trek. The beginning descent, at least, is rapidly accomplished. We find Peter, and the three of us reenter the open tableland with its deceptive grass and troublesome sandburs. Crossing the zone is like running a self-inflicted gauntlet. Our cardón sticks flail against our lower extremities. We want to move fast, and so we try to pick our way through spots where the ground is somewhat more barren. But now a new affliction arises: the burs adhere to the soles of our field boots. Once the first few burs attach themselves to our feet, they attract others by the score. In no time at all, the soles of our boots are dramatically thickened by up to an inch (2.5 cm) of densely pressed seedpods. Our bizarre tribe members hasten forward awkwardly, as if mounted on platform shoes. Every hundred yards (100 m), or so, we must stop and scrape the soles of our boots, one at a time, standing balanced on the other leg. Knowing how ridiculous this forced behavior is, we begin to laugh at ourselves. Once we leave the grasslands for the relative sanity of the palo adán thickets, the change in terrain is more forgiving.

Brief pauses are measured in small sips of water. When we cross the sidetrack 20 minutes away from the main road, my legs have all but forgotten the triumphant spring of certitude they previously felt with each outbound step. I take the last remaining swallow of water from my bottle. The uphill route to kilometer 41 on the gravel road is not steep but seems endless. My throat grows ever more parched, and my tongue feels as though it has no more space to expand within my dry mouth. Jorge has pulled the truck around to cast the greatest possible shadow in the afternoon sun. He hands each of us a bottle of energy drink, and we collapse, cross-legged on the ground, looking back at the great sand ramp. I glance at my watch: the time is

4:15 p.m. Our trophy, a small plastic bag filled with sand, required six hours of steady hiking to claim.

We sometimes lay fixed ideas on the landscape, wishing to confirm those patterns we most expect to dwell there. Features such as the great sand ramp at kilometer 41 exert a powerful attraction that calls out to anyone who will stop long enough to consider its simple riddle: "Why am I here?" Only days after returning to campus, Peter put the hard-won sand sample to the test under a microscope. The answer was definitive. There was no trace of carbonate sand to be found. Essentially, all the fine sand grains in the sample were siliceous in composition. Dave greeted this news as a special challenge and hunted down literature on similar sand constructions. He discovered that the term *sand ramp* was first applied to structures in the Mojave Desert of the southwestern United States by Landcaster and Tchakerian (2003). Essential to the concept is the idea that dunes can be topographically molded in settings where available relief in the landscape attempts to arrest wind-driven sand. The sand ramps of the Mojave Desert build structures ranging from 5 to 70 meters in thickness, typically manifested by eolian sand that deposits climbing dunes on the windward side of a fixed barrier and falling dunes on the leeward side of that barrier. The sand source comes from empty lakebeds, called pluvial lakes, that were filled with water thousands of years ago when glacial advancements displaced zones of precipitation that are normally found today in temperate latitudes around 45° to 55° in the Northern Hemisphere. It means that the Mojave Desert formerly enjoyed a wetter climate that came and went with the advance and retreat of Pleistocene glaciers on the North American continent.

By examining satellite images of the Baja California peninsula, Dave began to see evidence of additional sand ramps similar to the one we explored at kilometer 41. Good examples can be found near San Felipe and on the northern end of Isla Tiburón (Backus and Johnson 2009). The story makes a striking parable regarding the vicissitudes and ramifications of global weather patterns. Pluvial lakes did develop in some intermountain areas of Baja California. Today, their dry lakebeds occasionally become recharged with water under unusual conditions, as witnessed in Laguna Chapala (see plate 1), at the southern end of the Cataviña Boulder Field (see chapter 1). One can envisage the arid landscape of the Valle las Animas immediately north of kilometer 41 as cradling a set of interconnected lakes filled by water from the confluence of the San Pedro River and a stream draining part of the Valle las Flores. The sediments reaching these lakes would certainly include vast amounts of silica sand worn from the interior granite pluton of Sierra San Borja.

Sea level drops on a global basis during the episodic advance of continental glaciers, because of the transfer of ocean water to the atmosphere as part of the hydrological cycle. When precipitation falls as snow, it becomes semipermanently trapped as ice within great ice sheets. The ramifications for Baja California during glacial times included both a wetter climate and a reduction in the level of the Sea of Cortez by as much as 425 feet (130 m). The northern half of Isla Tiburón features a great, north-pointing valley some 18.5 miles long by 6 miles wide (30 km by 10 km), presently inscribing a watershed within the 650-foot (200-m) contour of topography. During a Northern Hemisphere glaciation, however, the entire Bahía Agua Dulce at the mouth of that valley would be exposed as land. Whether any of the carbonate sediments accumulating in that bay might be transferred landward to the valley's enormous sand ramp by stiff northerly winds is far from clear. It cannot be known for certain, until someone with this question in mind visits the place.

Meanwhile, our excursion to the great sand ramp at kilometer 41 remains stamped in my memory as one of the most fulfilling exploits I have instigated. We return to the apex of the ramp fairly often, not on foot but from above, comfortably seated in commercial jetliners in transit to or from places such as Loreto and La Paz. The spot where the climbing dune changes to a falling dune at the gap in the mountains is easy enough to pick out in the scrappy landscape below (see plate 4) as viewed from an altitude of about 30,000 feet (9,144 m). On recognizing the spot, I am reminded not so much of our day's trek but of the idea we tried to enlarge and the way we were challenged to grow by the experience.

When I choose to mentally revisit the place, I see things as spied not from the air or a position gazing out from the top of the ramp (see plate 3) but from my seat on the ground in the shade of the truck looking back up at the ramp. There, at that moment, I found a sense of deep satisfaction fed from a weary body and amplified by my heart's fulfillment of a goal long delayed. The feeling was one of sublime peace. In the everyday practice of our lives, it is easy to become agitated by the nuisances and disappointments that seem to conspire against us. They are not important in the greater scheme of things. At times when tempted to believe that such exasperations carry more weight than they deserve, I have only to recall the contentment of that moment. It returns, washing over me like a wave to gently pull me back to Earth. Kilometer 41, kilometer 41, kilometer 41, I repeat to myself like a mantra. It is my personal deliverance from the busy distractions of the conscientious traveler within.

4

San Francisquito's Ancient Bay

We made anchorage at San Francisquito Bay. The cove-like Bay is about one mile wide and points to the north. In the southern part of the bay there is a pretty little cove with a narrow entrance between two rocky points . . . and on the edge of the beach there was a poor Indian house.
—*John Steinbeck,* The Log from the Sea of Cortez

LISTENING TO OTHER TRAVELERS praise "Little Saint Francis Bay," I became eager to visit the place myself. The opportunity arrived in January 2002, when on that first occasion our group consisted of three professors and one student. We came by land, driving the roughly 80 miles (129 km) on dusty roads southeast from Bahía de los Angeles (see map 1, locality 4), not by sea as had John Steinbeck and Ed Ricketts aboard the *Western Flyer* in March 1940. On entering the inner north-oriented cove, Steinbeck described a keen sense of intruding on a very private place. An empty canoe sat beached in front of a lone, humble dwelling. The party quickly withdrew to another anchorage in the outer bay, where the next morning they sampled the local intertidal zone. True enough, San Francisquito remains today a secluded place. A few more cottages are in evidence at the end of the little cove. Beyond the cove to the south on the opposite side of low middle ground, something like a motel with a line of rustic cabins sits on a wide beach facing eastward onto the open gulf at Ensenada Blanca. Nearby, a dirt airstrip hints at a more immediate link with the outside world.

A single road reaches San Francisquito by crossing over a semicircular lip about 460 feet (140 m) in elevation before descending to the coast (map 4). By far, it provides the most commanding entrance to the place. Had Steinbeck and Ricketts traveled this way, the view may not have impressed them much. "The country hereabouts was stony and barren, and even the brush had thinned out,"[1] recorded Steinbeck, looking shoreward from the water. The coastal landscapes of Baja California often take on a double meaning for me, a significance that leaps out from the countryside and floods my mind with visions from another world. It is a sensibility that I doggedly labor to translate to my students. Whereas the common traveler sees only the contemporary coastline as viewed from sea or from a perch on land, I am conditioned to search for traces of an older shoreline now abandoned by the waves. In seeing the past imposed upon the present, such a double vision is not the kind of trick that conjures its magic automatically at any given summons. When it works, the revelation is like a gift. Beyond question, the scene that unfolds on the San Francisquito road when crossing the basin's rim is one of the most breathtaking sights I hold in my mental catalog of landscapes.

San Francisquito may embody other things to other visitors, but for me the landscape unfolds in exquisite beauty as a jewel of a Pliocene bay. It is an aged but oddly untouched geographic feature fully revealed in its nakedness by a dramatic change in sea level relative to those shores of former times no less than three million years ago. Certain aspects of this natural wonder are readily apparent: its size, its material boundaries, and a generous share of its exposed fossil treasure. Pliocene San Francisquito Bay is the largest found anywhere along the gulf coast of the Baja California peninsula. In size, it encompasses more than four square miles (10 km^2). The ancient bay is materially enclosed by granite. The bay's foreshore sediments, now expressed as conglomerate with a surfeit of cobbles and pebbles eroded from the adjacent granite coast, give ample evidence of the life that once thrived there. To cross the barriers of time and visit Pliocene San Francisquito is a dramatic experience.

How might the astonishment stirred by this remote spot be compared to the feelings engendered by other, more worldly sights?

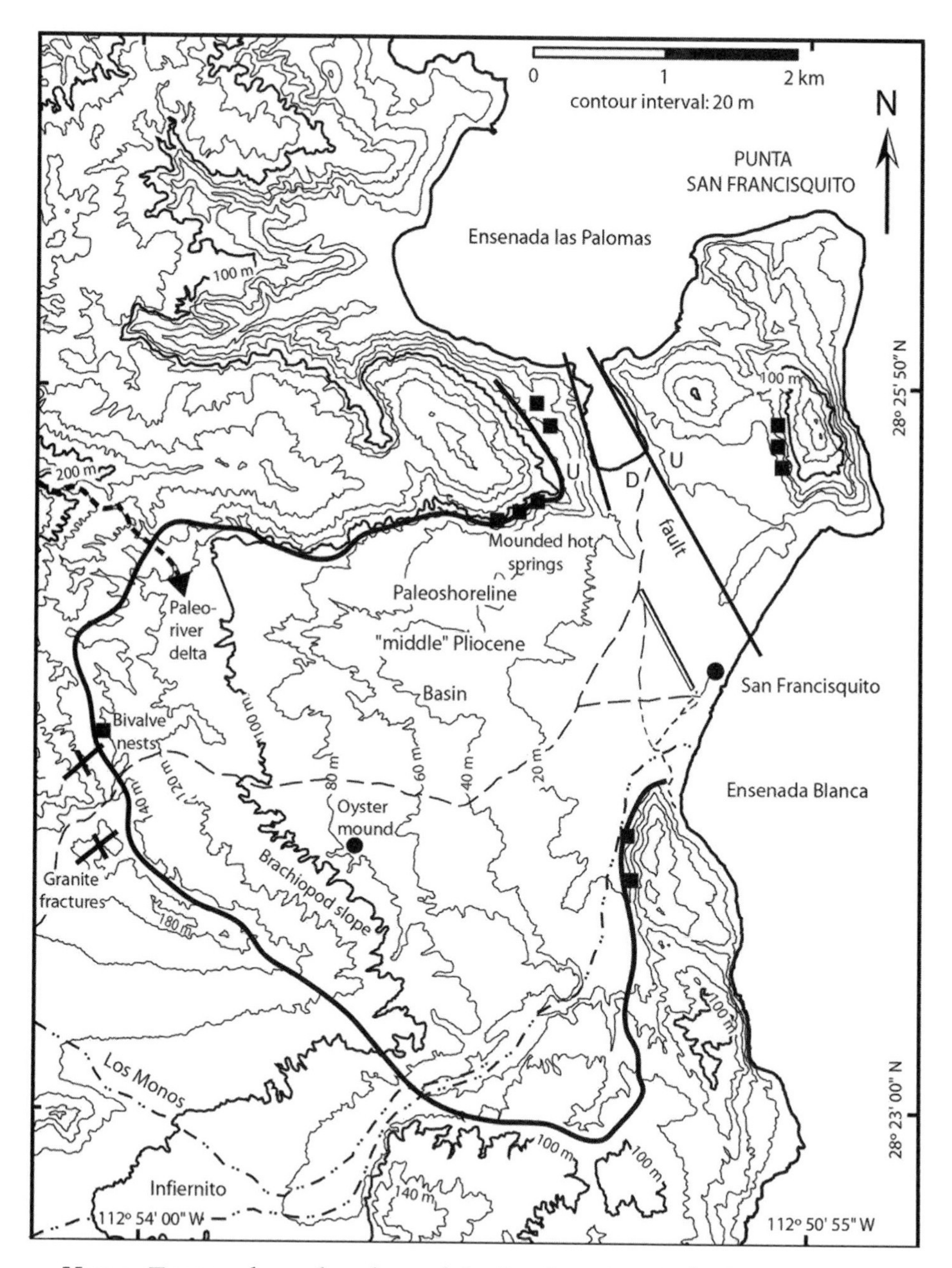

Map 4. Topography and geology of the San Francisquito basin. *Note:* Heavy black line marks a paleoshoreline; heavy dashed line shows the course of a paleoriver and location of a paleodelta; normal dashed lines show local roads; double line nearest San Francisquito marks an airstrip. Map by author.

To begin, granite adorns some of the most scenic rocky shores found anywhere on the planet. The Portuguese mariner Gonçalo Coelho led the first contingent of Europeans to enter the natural harbor at Rio de Janeiro on January 1, 1502. Not a river, in fact, but a commodious and elongated bay striking 20 miles (32 km) inland, the anchorage is guarded by granite monoliths of surrealistic size and shape. The most iconic is the 1,299-foot (396-m) Sugar Loaf (Pâo de Açúcar), which towers above its neighbor, Morro da Urca, a mere 722-foot (220-m) lump. During its colonial era, Hong Kong prospered as a strategic trading post, taken by British naval forces on January 20, 1841, and ceded back to China 156 years later in 1997. Ranks of granite crests enfold the natural harbors of this vibrant city. The most famous is Victoria Peak on Hong Kong Island, rising 1,811 feet (552 m) above the central city district.

Ascent by cable car to the heights of Sugar Loaf and Victoria Peak is standard fare for the tourists who flock to Rio de Janeiro and Hong Kong. Besides the impressive topography, the visitor is awed by the crowded neighborhoods that fill every available flat surface and encroach to an astonishing degree on surfaces that only can be described as vertiginous. Comparison of San Francisquito with these two great harbor cities, I realize, is a stretch of the imagination. The physical scale of landscape is all out of proportion, and the dry Pliocene bay is no recipe for a city bursting with human commerce.

Yet, it must be said that were the gracefully semicircular San Francisquito basin filled with seawater today to the same extent it once was (map 4), its attraction as a great harbor would be certain. In fact, such a propitious natural harbor would rival that of La Paz. Today, the La Paz harbor is the hub to a city of 215,000 inhabitants dwelling on the same great peninsula far to the southeast. The pleasant city *malecón* trails for a distance of two and a half miles (4 km) within the shelter of El Mogote, a massive sand and mud spit that protects the harbor from harsh winds out of the north during the winter season. By comparison, the natural arc of a Pliocene malecón around the inner San Francisquito bay would easily amount to more than six miles (10 km). To be sure, the granite hills surrounding the Pliocene bay were even more muted in topography three million years ago than they appear today. No steep city streets, as found in its near-namesake on San Francisco Bay in Alta California, were ever a prospect for San Francisquito.

Traces of the Pliocene shoreline remain in the small cove above its west wall. In theory, they might have caught the attention of Steinbeck and Ricketts as the *Western Flyer* turned about for a hasty exodus on March 31, 1940. Both men were experienced in marine biology, especially with the great panoply of invertebrates that inhabit the intertidal zone. Hailing from Pacific Grove on California's Monterey Peninsula, both were familiar with the kind of granite shores that embolden a seascape. Ricketts operated a supply house providing biological samples for teaching and laboratory use.[2] The granite tidal pools near Pacific Grove were a natural larder for his business. Steinbeck is renowned as a novelist, but he had taken a course in marine biology at the Hopkins Marine Station in Pacific Grove as part of his formal education at Stanford University. If noticed by the pair on that particular March evening, the fossil-rich limestone that fringes granite cliffs 160 feet (49 m) above the water's surface might have held the last rosy glow of the setting sun in the little cove. The hour was near 6 p.m., however, and it was getting dark. The *Western Flyer* left the inner cove and set anchor, instead, in the adjacent outer bay north of San Francisquito (Ensenada las Palomas).

The road to San Francisquito reaches an abrupt end at the cove abandoned by Steinbeck and Ricketts those many years past. During our first visit, we were fortunate to arrive while the day was still young. The morning sun shone brightly on the limestone cliffs above the cove. It required scarcely a half an hour to climb the slopes to the most advantageous spot where the limestone formed a low ramp dipping northward off the granite. Conglomerate composed of granite cobbles as large as three and a half inches (9 cm) is sandwiched between the limestone and the granite. Fossil shells of *Nodipecten arthriticus* are common in the overlying limestone layers. Many are disarticulated as single valves, and they typically bear encrusting barnacles (figure 12a). Steinbeck and Ricketts would have instantly recognized the meaning of these shells and associated barnacles as the mark of a higher sea level in former times. There was something else, however, that captured my attention at the outcrop and settled once and for all that I would make a return trip to San Francisquito for a longer stay.

Another sort of fossil is plentiful at the transition between the limestone and the granite cobbles. It is no marine invertebrate but

Figure 12. Pliocene fossils from San Francisquito: *a, Nodipecten arthriticus*; *b,* broken rhodolith (layered margin) encrusted on a granite pebble; *c–e,* brachiopods (*Laqueus* cf. *erythraeus*) showing pedicle valve, brachial valve, and shell commissure (side view); *f,* detail showing lacy bryozoans encrusted on pedicle valve of brachiopod in figure 12e; *g,* cluster of brachiopods in growth orientation; *h,* shark tooth. Photos by B. Gudveig Baarli.

a peculiar kind of calcified marine algae in the form of a rhodolith (literally "red rock"). Living beds of rhodoliths, or nonattached coralline red algae, make an important ecosystem in offshore settings found many places along the Baja California peninsula.[3] Belonging to the division Rhodophyta, these algae often grow in a spherical shape. They secrete calcium carbonate ($CaCO_3$) in concentric bands typically nucleated around a small piece of shell or a tiny fragment of a broken rhodolith. To carry out photosynthesis as plants, they must receive adequate sunlight through relatively shallow water. The shape they generally acquire is due to the fact that surface waves or gentle bottom currents frequently keep the spherical growths in sporadic circumrotary movement, which ensures more or less equal access by all surfaces to sunlight. In deeper, less wave-agitated waters, the rhodoliths also are turned over by the activities of marine invertebrates dwelling within the living beds (Marrack 1999).

Fossil rhodoliths entombed in the conglomerate/limestone deposit above the little cove at San Francisquito tell a somewhat more elaborate story. Each example is encrusted around a granite pebble, commonly one and a half inches (3.8 cm) in diameter (figure 12b). Given the close proximity of these fossils to the former shoreline, it stands to reason that the granite pebbles around which the rhodoliths nucleated in life were subject to frequent movement stimulated by the impact of ceaseless waves on an exposed shore. The waves must have been quite energetic to keep pebbles this size in sufficient movement to acquire a roughly even coating of coralline red algae with a crust up to half an inch (1.27 cm) thick. At the time of my initial visit to San Francisquito in 2002, I knew only one other example of a similar fossil deposit: rhodoliths nucleated around pebbles of eroded andesite on the shore of a Late Cretaceous island exposed in sea cliffs near the village of Eréndira on the present-day Pacific coast of Baja California (Johnson and Hayes 1993). Eréndira had made a deep impression on me as a place where a former seascape was vividly captured by the local geological record. Although it was a placid day at San Francisquito, my companions and I imagined that we stood in surf, perhaps waist deep, struggling to keep our footing against the onslaught of wind-driven waves during a Pliocene midwinter's day.

At a higher relative sea level some three million years ago, the same little cove at San Francisquito functioned as a through-going but narrow passage to the interior of a magnificent ancient bay from Ensenada las Palomas (map 4). Today, the granite hills forming the outer corner of the cove rise 525 feet (160 m) above sea level and provide a suitable lookout post for a Mexican army detachment on watch for drug runners. South from the lookout beyond the beach on which the San Francisquito motel resides, another ridge of granite makes a formidable rampart on the east side of the ancient bay. Between these two granite sentinels, the beach forms a gap one and a quarter miles (2 km) wide (see plate 5). When fully flooded during the Late Pliocene, the drowned beach became the main entrance to the huge bay located immediately behind it. Effectively, there had been enough space for 100 equivalents of the *Western Flyer* to enter the bay side by side from Ensenada Blanca. An entire fleet might easily rest at anchor inside the Pliocene bay, sheltered from unpredictable winter winds or other marine disturbances on the open Gulf of California. That would be something to behold, even if the granite heights surrounding the bay were vastly inferior in proportion to Sugar Loaf or Victoria Peak.

Because a larger crew was needed for a mapping party, the return visit to San Francisquito waited until January 2006 to become a reality. Our tent encampment was set up at the forlorn little motel, where we rented a drafty cabin mainly for the storage space it provided. We arrived on January 6 and enjoyed dinner that evening with the proprietors of the establishment on the enclosed veranda of the central "ranch house." An earthquake had rocked the place only two days before, we were told. The shaking was violent enough to send earthenware crashing from the shelves. The Gulf of California is the seat of relatively common but minor earthquakes caused by hiccups in the slippage of transform faults at shallow depths below the sea bottom. Preparing to sleep in my tent that evening, I wondered how I might respond if an aftershock caused the sand beneath me to undulate like a waterbed absorbing a sudden jolt. Perhaps I would react by calling out in alarm to my fellow campers. Surely, I would crawl out of my sleeping bag, grab my flashlight, and exit the tent as quickly as possible. It was not the prospect of the physical shaking that left me uneasy.

What about the possibility of a tsunami? Where we slept, our tents were situated less than 5 feet (1.5 m) above the high-tide line.

The thought of a possible deluge kept me from drifting off to sleep as rapidly as I might like. Only a little more than a year before, on December 26, 2004, a powerful tsunami was triggered by a massive earthquake off the west coast of Sumatra in Indonesia. The resulting devastation rippled with deadly force throughout much of the Indian Ocean basin, reaching as far as the Seychelles Islands off the east coast of Africa. More than 230,000 people lost their lives when waves up to 98 feet (30 m) in height inundated coastal communities (Paris et al. 2007). It was discomforting to think about such an aftermath in terms of the human toll. There is no historical record of such an event ever happening along the inner gulf coast. Instead of infrequent but extremely violent earthquakes, the region is known to absorb a rash of relatively low-energy events.

The overriding purpose of our January 2006 visit was to map the Pliocene San Francisquito basin with particular reference to variations in the distribution of marine invertebrate life left behind as fossils around the margins. As I finally drifted off to sleep, I wondered what discoveries awaited us in this remote place where few geologists and fewer paleontologists had previously ventured. At the same time, we could not lose sight of a related goal to discover how the basin had taken shape in the first place. The tectonic forces indigenous to the region surely had something to do with this process.

FEATURE EVENT

Walking the Natural Malecón on an Ancient Bay

Dawn creeps across the landscape on gentle fingertips under an open sky. With no clouds in sight, it appears that the sun will shine with an unwavering brilliance through the day. In the cool of the morning before anyone else rises for breakfast, I am drawn off the beach to the nearby dirt airstrip, where an unobstructed view across the southwest quarter of the Pliocene embayment is resplendent for its scale and geological context (figure 13). In the foreground is a stand of mighty cardón cacti worthy of acknowledgment as a *cardónel*, or desert forest. On the far horizon, the basin's granite rim hovers roughly 2 miles (3.25 km) in the distance. Like a curtain, a fringing mantle of conglomerate and limestone slopes forward, interrupted by several

Figure 13. View from landing field at San Francisquito to the southwest basin rim. Smooth low slopes below the horizon are formed by limestone sitting on granite. Photo by author.

gashes made by the erosion of deep gullies. Those clefts sit in dark shadows, while the granite massif that rises behind reflects a delicate rosy glow over the entire amphitheater. The upper slopes of sedimentary rocks forming the curtain are like a discontinuous bench, a wide and commodious malecón with a barely perceptible tilt toward the middle of the basin. Today, we walk the malecón over much of its circumference around this ancient bay, a hike of about 7 miles (11.25 km). It is not a contest to see who will be first to complete the circuit. We will walk together, not always slowly but with deliberate care to pick out all the interrelated nuances of past geography and ecology. The more eyes that are prepared to find interlocking pieces of the great puzzle, the better for us as a cooperative team. As I stand alone with the landscape, it is the sort of speech that I rehearse to rally the crew for the day's coming adventure.

Breakfast consumed and lunch packed for the hike, we set out on foot, following the motel's access road until it joins the main San Francisquito road (map 4, dashed line). Around a bend near the main road, an out-of-place reminder of our cultural zeitgeist makes an unexpected appearance. It is a bright-red banner bearing the silhouette of a Coca-Cola bottle in white together with the appropriate lettering pinned like a sail high on the trunk of an enormous cardón. The sign is the only reminder of civilization we can expect to see for the rest of the day. The main road leads southwest for six-tenths of 1 mile (1 km), before reaching an elevation 130 feet (40 m) above sea level and turning directly west. We are situated almost exactly in the middle of the ancient embayment. Continuing west for a short distance along the road, the frontal edge of the Pliocene escarpment appears as four principal lobes of strata separated by five large gullies among a confusion of lesser drainages. As soon as possible, we must abandon the road and strike overland through the desert bush, aiming for the southwest side of the basin. One of the larger gullies will provide a route to the basin's rim.

Here, the desert brush is dominated by nettlesome palo adán with its long, slender branches sporting stiff spines like sharp nails on a wiry rod. Progress is slow, but each step brings us slightly more upslope and closer to our goal. At the start of our off-road march, we spread out equally to make a frontal attack on the escarpment, each of us weaving on our own path between the tricky palo adán plants. Soon, however, the rank is broken, and the front member of the group drops into a dry streambed. The rest of us follow, now moving single file in the upstream direction. The arroyo is full of grus, the coarse by-product of weathered granite dominated by grains of clear silica and pink plagioclase. Here and there, however, a loose oyster shell litters the ground. The stray shells give us a clue as to what lies ahead.

After rounding a bend in the gulley, we arrive at a huge colony of fossil oysters that partially obstructs the streambed (map 4, oyster mound). Firmly cemented together as a mass, the colony is exposed in the west wall of the gulley to a height of 5 feet (1.5 m) and spills over the embankment to occupy a mound approximately 33 feet (10 m) in diameter. I carry a collapsible grid in my backpack that quickly can be assembled to make a sampling device used to estimate the number of fossils preserved in a natural population. The oysters are very abundant and represent a single species, *Ostrea fischeri*. Most of the shells are articulated and preserved in growth position. On average, some 23 individual oysters fill a square grid that measures 16 inches (0.5 m) on a side. The largest are 2.8 inches (7.1 cm) long and almost as broad at the widest part of the shell. However, the average oyster in the mound is a bit smaller, only 1.5 inches (3.8 cm) in length. Our discovery is significant, because it denotes the lowest elevation within the basin where marine life managed to colonize. Salinity may not have been normal during the initial phase of flooding, when the water depth was still shallow and the basin's connections to the open gulf were perhaps more restricted. Without excavating further around the edge of the oyster mound, we cannot know exactly what kind of surface the colony first established itself upon. Most likely, however, it was a knob of granite that projected somewhat above the basin floor.

Today's mission is a reconnaissance of the basin margin. We cannot afford to tarry long at any one place. As we continue upstream through the arroyo, limestone strata with a distinct admixture of coarse silica sand are found to bury the oyster mound. Pecten shells are the primary fossils at this level. A quick check reveals that two species are present: *Argopecten antonitaensis* and *A. revellei*. These species, both now extinct, confirm a middle to later Pliocene age for the basin based on their stratigraphic range as index fossils in the thick pile of sedimentary rocks left along parts of Baja California's gulf coast.[4] The elevation of the basin floor at this spot is roughly 230 feet (70 m) above sea level. Ahead, the prospect of the middle gulley provides a route upward through a steeper part of the escarpment to reach an outer ledge about 33 feet (10 m) above our position. We enter the gulley single file but soon become stalled in a tangle of desert brush. The only choice is to scale the sidewall on our flank. It is a slow, sweaty business that quickly leaves me winded. On reaching the cliff edge, however, we are rewarded by open views back over the basin's center—and more enticingly, forward over a gentle gradient rising to intersect the basin's natural rim.

Nothing attracts my eye by way of eroded fossils on the ground. It will require a climb through almost another 200 feet (60 m) to reach the unconformity between the granite and the fringing basal conglomerate of the Pliocene sedimentary sequence. I am eager to get there, and I charge ahead. The next part of the incline breezes past in a flash, and I am soon beyond the 330-foot (100-m) contour. Dave calls out to me to stop. With eyes fixed on the ground, he has taken his time coming up the slope and has found something. "Get back here!" he yells. "We've got brachiopods." In my eagerness to reach the unconformity, I have violated my own admonition against racing. Brachiopods? We had found them only once before, when exploring the south side of the San Nicolás basin near Punta San Antonio, far to the southeast in Baja California Sur. As I retreat down the slope, I recall the handful of delicate, disassociated brachiopod shells we had collected from the floor of a washed-out gulley. They were scarce, and we never managed to find the rock layers those shells derived from. None of the field guides or larger treatises on invertebrates in the Gulf of California mentions hinged brachiopods from the three surviving orders of the once-thriving phylum, the Rhynchonellida, Terebratulida, and the Thecideida. We had found fossil scraps of terebratulid brachiopods near Punta San Antonio, possibly *Laqueus erythraeus.*

Brachiopods are among the most abundant fossils in Paleozoic rocks ranging from the Ordovician through Permian systems. When my students accompany me on local field trips to the Devonian Helderberg escarpment near Albany, New York, they are guaranteed to find as many brachiopod fossils as they would care to collect. The same can be said for the Ordovician rocks around Cincinnati, Ohio, or the Silurian rocks around my boyhood home of Dubuque, Iowa. Brachiopods (literally "arm-foot," from Latin roots) show a superficial resemblance to bivalves, or clams. However, most of the clams we are familiar with possess two valves that are the mirror image of one another. Brachiopods do possess two shells, but one is larger than the other, and it usually reveals a small opening near the hinge through which a fleshy stalk protrudes in life to tether the animal on the sea floor. The fleshy stalk is called a pedicle (the "arm-like" foot), and the valve it is associated with is the pedicle valve. The smaller of the two shells is called the brachial valve, and it typically includes internal structures that act as the skeletal support for the animal's feeding appendage. Thus, the two shells of the brachiopod are dissimilar from one another because they serve different functions.

As I reach Dave, he shows me a handful of small brachiopods that are fully articulated, with cojoined pedicle and brachial valves (figure 12c, d). In side view (figure 12e), the larger valve is readily apparent from the beak-like protrusion that extends beyond the hinge area. Like the shells from Punta San Antonio, these are terebratulid brachiopods in the genus *Laqueus*. There are thousands of them scattered around on the ground, fully weathered out from the Pliocene rocks that once entombed them. It is an astonishing sight, one that feels oddly out of place for an embayment filled with Pliocene strata. We begin to comb the area in a lateral direction parallel to the bedding exposed in the escarpment. We spread out along the hill's 330-foot (100-m) contour line. Virtually all of them intact, the shells appear to occur within a span of 26 feet (8 m) up or down the incline. They are present at this level across the entire sweep of the central lobe on the escarpment.

The first specimens I pick up are fine enough, but soon I become choosey and save my pockets for the best. Two specimens, in particular, are prize discoveries. They are plump individuals, with a slightly longer shell length compared to shell width than most of the other brachiopods. Each has a small hole about three-eighths of an inch (1 cm) in diameter poked through the eggshell-thin casing on the brachial valve. Looking inside one and then the other, I see that the bottom part of each brachiopod interior is filled with lithified sediment. This means that the lower parts of these specimens are solid like a rock but the upper parts are hollow. This is an artifact of preservation, called a geopetal. It shows that the brachiopods stood erect on their short pedicles, rather than lying prostrate on the sea bottom with the brachial valve uppermost.

Under strong sunshine, I cup a hand over the broken opening on one of the shells to see more clearly inside. I can scarcely believe my eyes. Fragile skeletal structures are exposed on the inside of the brachial valve. I see parts of the looped brachidia, or small hoops, that supported the lophophore, or feeding appendage. I am instantly carried back to a summer's morning on the west coast of San Juan Island in the upper reaches of Puget Sound near the border between Washington State and British Columbia. It was early in the morning, and the lowest tide associated with the lunar cycle had reached its nadir. I was taking a graduate course in marine invertebrate biology at the Friday Harbor Marine Laboratory in 1974. Our instructor, Eugene Kosloff, showed us how to find living brachiopods under flat stones exposed at tide level. I had recently completed the first year of graduate studies as a budding paleontologist, but these were the first living brachiopods I had ever seen. I had always viewed them as fossils, cast only in ghostly shades of gray. The living terebratulids of Puget Sound possess shells that are shockingly pink. The middle Paleozoic was the era of the hinged brachiopod, but today members of the phylum are rarely seen at the seashore. Back at the lab later that day, we dissected some of the brachiopods. I learned that they pack scarcely any meat, particularly in comparison to bivalve mollusks. For a predator such as a starfish seeking a good meal, opening a brachiopod shell could cost more energy than might be gained from consuming the inner tissues. Unlike mussels, for example, brachiopods would hardly seem worth preying on. Perhaps that was the secret of their Paleozoic success. At San Francisquito, however, brachiopods did host other organisms such as lacy bryozoans that encrusted their shells (figure 12f).

Moving slowly up the incline of the Pliocene basin with a renewed sense of focus, I spy something extraordinary. It is a cluster of four brachiopods preserved together in original growth position (figure 12g). The cluster of shelly material had worked free from a rock layer and sat, now, upside-down with the posterior part of the shells pointing up at me. Here is another line of evidence regarding the life habits of these brachiopods consistent with the geopetal fillings observed inside half-empty shells. The thousands of brachiopods that once thrived in relatively shallow water off the shores of the great bay did so in such crowed proximity that many of the individuals grew upright on their narrow hinge area, supported by the shells of the surrounding neighbors packed around them.

Many brachiopods became extinct by the end of the Permian Period with the close of the Paleozoic Era. Molluscan bivalves remained in the shadows during the Paleozoic primacy of brachiopods. After the Permian mass extinctions, however, the bivalves largely replaced brachiopods and took over the Mesozoic world as a dominant marine invertebrate (Fraiser and Bottjer 2007). Yet other revolutions occurred after the end of the Cretaceous Period with the close of the Mesozoic Era, when some bivalves adopted a successful infaunal habit (living below the sediment surface) to avoid predation (Stanley 1968). The common Venus and razor varieties, emblematic of these later tribes, are well known to those who dig for clams. I cannot help but feel a sense of awe at the utter fecundity of the Pliocene brachiopod populations at San Francisquito. The bad news about mass extinctions and the decline of the brachiopod world never reached these parts. In more ways than one, the embayment at San Francisquito was a Pliocene haven for life that found refuge here.

Somewhere above an elevation of about 400 feet (122 m), we reach the first pavement of the basin's natural malecón, where the younger limestone layers thin to a thickness of 16 inches (40 cm) and can be viewed along strike sitting on a thin mantle of conglomerate, in turn resting on granite basement rocks. In a cross-section view incised by one of the persistent gullies, the limestone looks something like a great wedge that thickens toward the center of the basin while conforming to the physiographic ramp. The noon hour is not far off, and we hasten to the northwest, climbing another 60 feet (18 m) to the rim of the basin, where some patches of limestone are still evident. The walking is easy, and after about a mile (1.6 km), we encounter the San Francisquito road. Here, the track is nestled in a narrow passage formed by a loop in the 460-foot (140-m) contour (map 4). The sun is at a high angle, but we find a stretch of the road pointed in a northerly direction. Crouching together, we find some measure of shade below cliffs at the side of the road. It is a good place to pause and take lunch.

I am impatient to see more of the contact between the Pliocene limestone and underlying granite. The defile in which the road is slotted conveniently opens below onto an area where the landscape reveals an uneven surface in naked detail. Several cardón cacti are growing from thin soil on granite (see plate 6). They lack the same stature as the giants in the cardónel closer to the airstrip but are healthy. Here, the tallest individuals rise perhaps eight feet in height (2.5 m). The erect tips of the cacti fail to rise above the level of the nearby gray limestone. In the distance, a ridge of granite emerges from beneath the limestone and runs sideways in a direction oblique to my line of sight. The ridge is exhumed from the limestone but draped by an overcoat of loose granite boulders. The complexity of the basin is unexpected. It is nothing like a simple bowl with smooth sides angled toward the center at a uniform rate of decline. Rather, it shows a crudely corrugated surface that undulates downward toward the bottom of the depression. The basin is more subtly textured in design than might be first imagined. In effect, the rippled topography of the granite allowed a succession of small lagoons to pool behind low barriers as the level of the sea advanced over the landscape. I am strangely attracted to this peculiar tableau, but the dusty floor at the feet of the impassive cacti yields no additional secrets. I must content myself by climbing up through the layers of limestone that define the nearby edge of the escarpment, and I return to the road by a circuitous route. Dave and the students abandon the scant shade at the cliff side and come to meet me.

There is work to be done here, where the limestone thins to the west along both sides of the road. To start, this is a good spot to measure the dip angle of the limestone where it impinges on the road. The limestone layers make an incline of 5° to the northeast. As we walk along the south side of the road toward the rear of the basin, the limestone vanishes to expose the underlying basal conglomerate. A veneer of granite cobbles and pebbles in a matrix of silica sand remains fixed in place, as if washed by the last sweep of an incoming wave. From my pack, out comes the collapsible grid with its meshwork of strings marked off in squares measuring 3.9 × 3.9 inches (10 × 10 cm). A random sample of smooth, pink granite cobbles is framed. They are elliptical in outline, on average 2.5 × 5 inches (6.5 × 13 cm) in dimensions. The cobbles grade to an assortment of mixed large, white quartz and pink granite pebbles, on average 1.5 × 2.4 inches (3.75 × 6 cm) in dimensions. At this locality, the embayment's shore was showered with wave-sorted clasts eroded from the parent granite. The white pebbles are polished bits of quartz that come from original pegmatitic dikes in the granite or possibly from secondary quartz veins. The latter would suggest hydrothermal action, but such activity was not necessarily ongoing during the formation of the bay.

As we continue beyond the conglomerate to the bare basement rocks behind, a distinct network of crossing joints is revealed. These more or less vertical fractures in the granite are not equally spaced. The principal joints trend in a northeastern-southwestern orientation and demarcate parallel spans of granite that range from 14 inches (44 cm) to 25.5 inches (80 cm) in width. Secondary joints follow a northwest-southeast orientation to incise subparallel spans of granite 4.5 inches (15 cm) to 20.5 inches (64 cm) in width. The joints do not cross at right angles but meet to form open obtuse angles of about 100° with an east-west orientation (map 4). It means that any waves striking the back of the basin from the east would act to quarry blocks of granite having starting dimensions with a planar ratio of approximately three to one. As suggested by clast sizes preserved in the nearby basal conglomerate, the resulting granite cobbles were reduced by wave action to dimensions with a planar ratio of nearer two to one.

Changing course to recross the road, we continue north on bare granite for some 500 feet (152 m). Here, the gross pattern of primary and secondary joints is found to duplicate that observed south of the road (map 4, dashed line). It begins to dawn on me that perhaps some of the topographic irregularities in the granite basement rocks observed as ridges exhumed from beneath the limestone farther out in the basin are related to this system of joints and the variations in their spacing. Orientation of the ridges seems more closely parallel to the trend of the secondary joints, but if so, the ridges are more massive while the intervening troughs are dissected by a greater number of joints.

Except for a few ghostly outlines shaped like bivalves, fossils had been difficult to find in the limestone near the road. Now, we are back in business on a limestone ledge 60 feet (18 m) wide at an elevation 480 feet (146 m) above sea level. In no time at all, we manage to identify the internal molds characteristic of four bivalves (*Glycymeris maculata*, *Anadara multicostata*, *Cardita megastrophia*, and *Cardita affinis*). These are chiefly surface-dwelling clams that have a habit of nestling together. The fossil bittersweet clams (*G. maculata*) are about 2 inches (5 cm) across and rather abundant. All the fossils at this site are whole (articulated) and appear to be preserved in growth position. This type of clam has a strong toxodont hinge with a straight line of small "teeth" on one valve that slot into matching holes on the opposite valve. Once again, the portable grid is brought out of my pack. A small population of 10 bittersweet clams is found to occupy a flat, rectangular surface measuring 20 inches (0.5 m) on a side. At another random position but farther out on the ledge, a population of 22 whole bivalves crowds the grid. Today in the Gulf of California, members of the family Glycymeridae typically live just offshore in shallow water.[5] Inside the three-million-year-old Pliocene embayment, that is precisely where we are ecologically situated. Further scrutiny of the limestone ledge finds additional denizens, including the shallow-water gastropods *Strombus subgracilior* and an unidentified species belonging to *Conus*. Many living conches (family

Strombidae) are intertidal in habit.[6] A more refined ecological picture of the ancient bay is emerging.

Encouraged by the discovery of the fossil gastropods, I wish to keep close to the paleoshoreline demarcated by the limestone's basal conglomerate. We continue in a northwesterly direction moving gradually upslope but angled toward the granite coast. After barely a quarter of a mile (0.4 km), we reach a place where a single granite boulder with a circumference of about 40 inches (1.25 m) rises above the surface like a mound (figure 14). The boulder is capped by 10 inches (25 cm) of limestone stained with a crust of dark minerals, possibly ore of manganite. Viewed in cross section, the dark limestone layer includes the molds of a dozen articulated bivalves, most likely bittersweet clams, given the size and profile. The limestone filling narrow cracks between smaller boulders on the periphery is similarly discolored. We seem to have discovered a "fossil" hot spring in the very shallow-water to intertidal zone along the paleoshoreline. It is the first evidence that the kind of thermal activity associated with secondary quartz veins in the granite was a continuing affair contemporaneous with the Pliocene flooding of the basin.

A few steps way, we discover a much larger mound of granite boulders with darkly stained limestone filling spaces between the igneous clasts. Here, again, the limestone holds the rough molds of many articulated bivalves. The mound is 23 feet (7 m) in diameter and rises about 3.25 feet (1 m) above the existing surface. A somewhat larger but very similar-looking mound is sighted to the northwest. We pace off the distance and arrive at the next mound after traversing a little more than 100 feet (31 m). The same darkly stained limestone is present as a thin cover over granite boulders. As before, the limestone is filled with the molds of fossil bivalves. The exercise is repeated, yet again, as a third mound is sighted in a direct line with the previous two. Together, the three mounds are arrayed along a straight path stretching for 525 feet (160 m) on a bearing of N 45° W (figure 14). Considering the fact that *G. maculata* is the most abundant bivalve fossil identified along the limestone malecón, it is reasonable to think that many of the molds concentrated in the various mounds represent the same species of bittersweet clam (figure 14, inset drawing). The mollusks appear to have thrived in raised rock gardens bathed by the warm water from thermal springs.

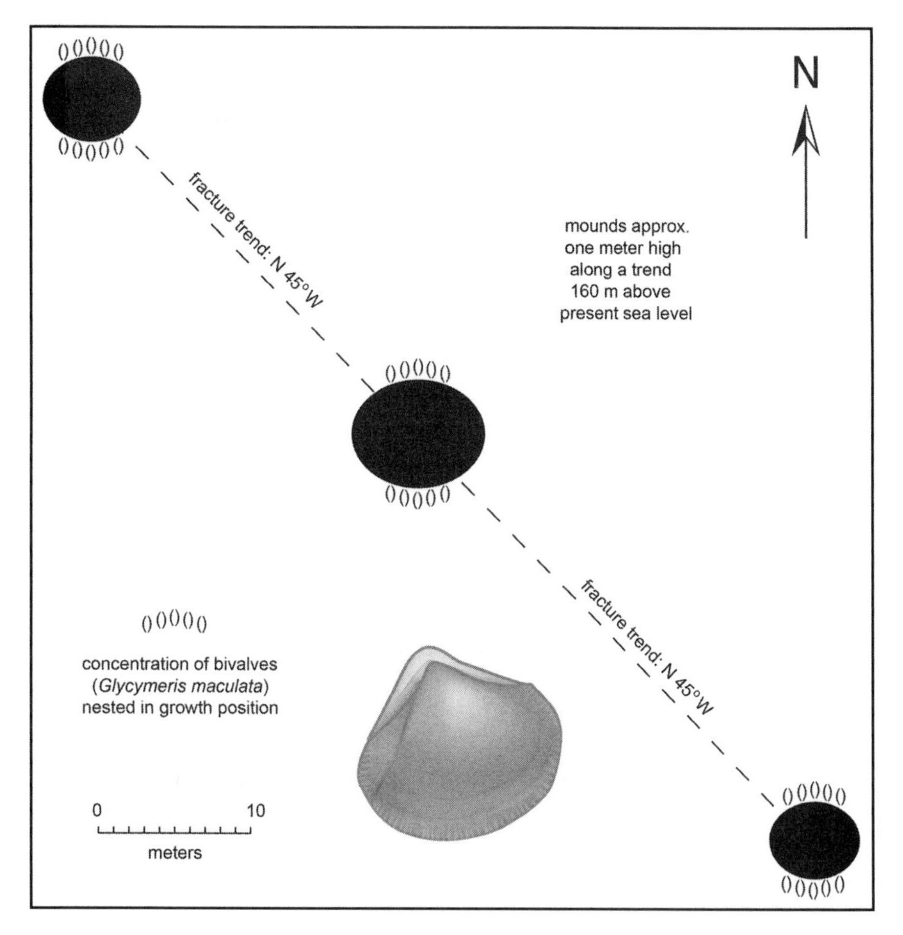

Figure 14. Field sketch showing relationship between mounds with a thin cover of limestone stained by dark minerals that denote a former hot spring colonized by marine bivalves (*Glycymeris maculata*). Original drawing by B. Gudveig Baarli.

Shallow-water hydrothermal springs are known from present-day sites along the shores of Bahía Concepción about 125 miles (200 km) southeast of San Francisquito (Forrest et al. 2005). These active springs sit in waters up to 43 feet (13 m) deep and with temperatures as high as 198°F (92°C). Such a high temperature is too extreme for many marine organisms such as fixed bivalves, although mobile gastropods belonging to the genus *Nassarius* are attracted to the vent sites. Thus, it seems likely that temperatures were more moderate at the San

Francisquito hydrothermal springs, which were perhaps regulated by tidal conditions. The Bahía Concepción vents are aligned over a linear distance of nearly a mile (750 m) roughly parallel to El Requeson Fault on the west side of Bahía Concepción. El Requeson Fault, itself, trends N 30° W. That is the same average trend for secondary joints in granite at the rear of the San Francisquito basin. In contrast, fracture zones associated with transform faults in the present-day Gulf of California commonly trend on a bearing around N 45° W. To what extent are these orientations coincidental? Whatever their precise origin, the "fossil" hydrothermal springs along the inside margin of the Pliocene bay combine aspects of local ecology with local tectonics and rock mechanics.

From our position near the 525-foot (160-m) contour, the basin's edge appears to take a corner on our intended circuit. Descending slightly in elevation, we continue in a northwesterly direction, spotting no further traces of former hot springs. Soon enough, however, the landscape's natural curvature has us change course to follow a more northeasterly direction (map 4). Dropping lower in elevation, we arrive at the mouth of a canyon that opens abruptly onto the basin from the northwest. Above this junction, the south wall of the canyon is 18 feet (5.5 m) high. A distinctive conglomerate layer composed of small granite boulders with an overall thickness of 3.25 feet (1 m) caps the canyon sequence. It is nothing like the shore conglomerates to which we have grown accustomed on our march around the basin. The clasts are elongated, just as we have seen before, but they are larger in size and all are in direct contact with one another (figure 15). Furthermore, the clasts are packed together to show a preferred orientation. *Orthoconglomerate* is the technical term applied to this kind of deposit. These boulders are crudely imbricated, or aligned to overlap with one another. Here we find all the hallmarks of turbulent, unidirectional flow in a river under flood conditions.

Consulting the topographic map, we can plainly see that the present canyon leads inland, where it is entirely confined by solid granite walls. This corner of the landscape has changed hardly at all in three million years. As a rare event, a tropical storm over this region could dump enough rainfall to create a gully-washer. Such a storm would flush the arroyos above us and transport a torrent of stones downstream. At the thought of so much water, impulsively, I reach for

Figure 15. Conglomerate composed of oriented granite boulders that represent a river deposit under flood conditions (scale represents 5 in. or 15 cm). Photo by author.

my water bottle and take a long drink. In this arid setting, we stand nearly in disbelief as witnesses to a bygone flood. The thick layer in the canyon wall immediately below the conglomerate is composed of coarse grains of silica and plagioclase. The buff sandstone looks remarkably like the unconsolidated grus in the present streambed of the arroyo. Beneath the sandstone is a thick layer of red siltstone penetrated by numerous burrows in all manner of orientations. Silt was deposited at the edge of the bay during the waning stages of a flood, making a kind of river-mouth deposit where shore crabs lived (map 4, paleoriver delta).

We cross the arroyo on a diagonal path to reach the opposite cliff face, similar in height. The cap rock is formed by sandy limestone, nearly five feet (1.5 m) thick, with traces of articulated bivalves exposed in cross section. Below the limestone is a layer of orthoconglomerate, essentially contiguous with the conglomerate unit now behind us but thicker by 50 percent. Beneath the conglomerate is more buff sandstone. Like its counterpart on the opposite side of the canyon, the sandstone is composed of washed grus, but with granite pebbles scattered through the upper part. The lower part features laminations one-sixteenth of an inch (1.5 mm) in thickness with climbing ripples. That is, thin sheets of sandstone preserve fine ripples superimposed on one another such that the crests of successive ripples are offset against those immediately below. Experimental studies using laboratory flumes show that climbing ripples typically aggrade when the sediment load carried by a stream increases. Strata here indicate that fluvial materials imported into the bay by floods of varying magnitude were eventually drowned during a relative rise in sea level that imposed a limestone cap.

Moving southeast, still above the 400-foot (120-m) contour, we encounter strata that tell yet another variation in the chronicle of the Pliocene embayment. The same kind of sandy limestone we last visited caps off the succession. Below is an enormous heap of granitic conglomerate exceeding 21 feet (6.5 m) in thickness. It is not an orthoconglomerate, as such, because the granite cobbles are awash in sandy material. If fossils are present, they are exceedingly rare. Beneath the conglomerate is a more respectable conglomerate, 6.5 feet (2 m) thick, composed of granite boulders up to 28 inches (70 cm) in diameter held together by sandy limestone. The matrix includes abundant fossil molds of bittersweet clams (*G. maculata*) nestled together in growth position among the boulders and cobbles. This layer sits atop a solid basement of granite. The contact between the granite and overlying conglomerate gives us the most evocative portrait of a former rocky shore that we have encountered all day. The rocks are bold and emphatic in their story. We are approximately 3.5 miles (5.75 km) west of the main passage into the bay from the open Gulf of California. Perhaps infrequent, any easterly winds pushing waves directly into the bay would find their greatest impact here at the west corner. Granite clasts in the overlying unit may have been reworked from imported river material or derived from the granite ridge flanking us to the north. Either way, the limestone cap reconfirms burial under a relative rise in sea level. The whole package dips 6° to the southeast.

As we approach the 330-foot (100-m) contour across the basin, it is clear that the escarpment to the north forms the steepest section of the basin rim (map 4). It maintains a bold presence on our left as we continue to descend down the slope into the basin. We are no longer on the malecón but below it looking up through about 33 feet (10 m) of strata composed of limestone layers atop a wall of washed grus sandstone containing scattered pebbles of granite and disarticulated pecten shells. The beds dip south, toward us, at what seems to be an unusually high angle in excess of 15°. We skirt along the lower beds with the pecten shells. Here and there, a few granite boulders with a diameter of 20 inches (0.5 m) are fixed in the sandstone as if floating. They must have tumbled down from the basin's rim in a rockfall. Something flashes from the sandstone against the bright sunshine, and I bend to examine it. It is the dentine of a shark's tooth. The blade is scarcely 0.8 inches (2 cm) long, but with serrated edges like a steak knife (figure 12h). It was a smaller animal (*Carcharodon carcharias*), in the ancestral line leading to the great white shark.

The sun sinks lower behind us in the west, and there is no time to tarry. We seek an accessible path leading up through the north escarpment, but the basin rim is like a formidable rampart. In another mile (1.6 km) on our eastward march, the escarpment is less daunting. As we press forward, a line of three "fossil" hot springs with the same characteristic dark staining in limestone appears on our flank. The lineup of these features does not match the orientation measured in the other "fossil" springs we visited after lunch (see figure 14). It is more likely that the springs on this segment of the escarpment (map 4, mounded hot springs) are related to fractures below the limestone that parallel the principal joints in the basement granite.

At the far northeast corner of the basin, not much inland from the cove visited by the *Western Flyer* in 1940, we find the path up the escarpment that we have been searching for. A cut in the strata allows us to scramble up 190 feet (58 m) along the sloping limestone to an elevation about 312 feet (95 m) above sea level, where the unconformity between granite and limestone is exposed. Unlike the sandy limestone we have grown accustomed to, here we find limestone in which the coarse products of weathered granite are far less pervasive and very fine sand is a component. Fossils consist of broken shell fragments from small pectens and the scattered spines of echinoids. More intriguing, the dip angles preserved in the limestone beds steepen upslope. Layers lower in the escarpment are dipping 20° to the southeast, while higher in the section the overlying limestone beds are dipping about 25° in the same direction. Because the beds are conformable, showing no obvious break or disjunctions in the succession, it means that the ramp slowly became steeper as the limestone was deposited.

Changes in the dip angles of limestone formations around the periphery of the basin and the discovery of "fossil" hot springs that align along the same margins give us much to think about. The view over the San Francisquito basin from our perch on the corner of the north escarpment is dizzying, both for the gapping void spread out below us and for the muddle of accumulated facts now pressed upon us from the day's hike. Dusk will descend before too long, and the north end of the dirt airstrip is more than a half mile (1 km) away. It is time to return to camp.

On the morning of April 1, 1940, Steinbeck and Ricketts went ashore to explore the intertidal zone around Ensenada las Palomas adjacent to San Francisquito.[7] The embayment is smaller and far more exposed to the wind than is the Pliocene bay hidden next door. The tide had dropped two feet (0.75 m) overnight. They reported finding barnacles, limpets, two species of snails, three kinds of crabs, and many large chitons. Lower down beneath rocks, they found great masses of a tubeworm with rusty red gills, some tunicates, and holothurians (sea cucumbers). Later that day, they baited nets and set them out to see what bottom life might exist. When the nets were pulled aboard the *Western Flyer*, they were startled to discover that one of the nets had captured a horn shark (*Gyropleurodus*, now referred to the genus *Heterodontus*).

During our ensuing days on the project in January 2006, the crew explored additional avenues on land to complete the San Francisquito story. The little north cove shared more of its secrets. For example, its flanks are bound by faults that define a small graben, and additional "fossil" hot springs are found seated in the limestone above the west side of the cove. Still more "fossil" hot springs are located along a fault east of the cove (map 4). The confluence of the Infiernito and Los Monos arroyos along the inside east margin of the Pliocene embayment leads to still more hot springs and the discovery of fine-grained, sandy limestone rich in echinoids.

Much camp discussion was devoted to finding a reasonable explanation for why the Pliocene embayment originated and subsequently evolved as a landscape. Key to our thinking was the fact that the area's basement granite clearly expressed primary and secondary joints defining obtuse angles open to the east. Wind-driven waves entering the basin from the east retained the most energy to pluck out individual stones from the rocky shore sliced by these joints. The waves rounded rough stones into smooth boulders, cobbles, and pebbles. Much silica and plagioclase sand was introduced by not one but two river systems emptying into the northwest and southwest quarters of the basin. Even more significant, the area was shot through with hydrothermal springs that locally weakened the granite and promoted its disintegration. The graben forming the small cove and the fracture along the embayment's inner west shore mimic fault trends still common in the Gulf of California today. Acting against other contrary faults within the basin, any lateral movements along northwest-southeast-trending faults during the later Pliocene slowly squeezed the basin like a hinged rhomboid (Shepard 1950). The resulting mechanics thereby deformed and oversteepened some of the sedimentary ramps growing inside the bay.

In the end, San Francisquito was mentally revived as the grand, water-filled embayment it once was three million years ago. Like city districts arrayed around the arc of a natural harbor, the inhabitants are divided into discrete enclaves. Echinoids plow through the fine sediment inside the highly sheltered northeast and eastern flanks of the basin. Brachiopods throng the basin's southwestern ramps in a shallow, subtidal setting. Mollusks dominated by bivalves affiliated with the bittersweet tribe are concentrated on the western flank of the basin, where they live offshore on a shallow bank, but also within the coastal zone nestled among granite boulders. Oysters stand at the ready to take advantage of any newly available rocky prominence. Sharks enter to troll inside the bay. Rhodoliths roll back and forth only outside the basin around the restricted north passage into the bay, where waves are turbulent much of the time. Lonely ghosts hang tenaciously onto this forgotten landscape but embrace the visitor seldom to these parts who would stay and hear their tales.

5

Lost Lagoons of Bahía Concepción

During the late Pliocene some parts of the mapped area were downwarped enough to be covered by the waters of the Gulf with resultant deposition of a thin veneer of sands, marls, and coquinas, forming the Infierno (?) Formation.

—*C. Carew McFall,* Reconnaissance Geology of the Concepcion Bay Area

SEEMINGLY INERT, A LANDSCAPE truly breathes under its own history of change and ongoing modification. Among those most conscious of the living land, geologists cultivate a strong pictorial sense of their art. Anything well illustrated is easier to understand than by text alone. When a field study is concluded, one's experience with a specific three-dimensional tract of countryside must be conveyed to an audience as effectively as possible. With fossils and the notion of time added to the blend, we are constantly challenged to look, think, and communicate about the Earth in four dimensions. Hence, it is customary for geologists to use serial cartoons to explain how landscapes evolve. What could be more direct than a kind of film clip composed of multiple frames shown in sequence to animate a developing landscape?

In this context, the geological map is the ultimate stop-action frame that depicts a given terrain, typically in its present state. Some maps are so subtle that intricate details wordlessly speak volumes about how a landscape emerged through time. As a corollary to the rule that a good geological map may surpass the written word, some are so powerful they compel the map reader to go visit the place in person. Such a map may speak a dialect unfamiliar to the mapmaker, because it accurately portrays relationships precluded or perhaps not fully addressed by the accompanying text. The printed word is limited and intractable, but a good map may transmit information not fully appreciated by its author. Thus, a geological map may speak differently to different people, because the potential audience includes members with diverse interests and agendas.

Rarely have I been more captivated by a map than I was by the 1968 reconnaissance map produced by C. Carew McFall showing the geology of the region surrounding Bahía Concepción south of Punta Chivato in Baja California Sur (see map 1, locality 5). During the mid-1990s, Punta Chivato was the place where some of my students cut their teeth on the extraordinary combination of geological and paleontological phenomena available for study along the gulf coast of the Baja California peninsula. The Punta Chivato landscape is singular for the way it projects itself crossways (as an *atravesada*, in Spanish) into the Sea of Cortez (see map 1). From almost anywhere on the south slopes of its promontory, the view over Bahía Concepción is utterly breathtaking. Jumping over Punta Chivato's "fossil" islands in this accounting is difficult,[1] but from its vantage point, Bahía Concepción and the great arm of the Concepción Peninsula loom large on the horizon.

The grand scale of the bay is unique to the gulf coast. To quote McFall (1968, 2): "Concepcion Bay is 23 miles [37 km] long, 2 to 3 miles [3.22 to 4.83 km] wide, and parallels the Gulf of California, with which it is connected at the north." McFall was a research associate at Stanford University's School of Earth Sciences, and his foray to the Concepción region was undertaken long before the Mexican federal government opened the paved highway from Tijuana to La Paz in 1971. Prior to this historic achievement, the overland route required weeks of travel from end to end. Segments of the old road remain visible from the modern highway as relics chiseled into cliffs along the west side of Bahía Concepción. In McFall's day, access to the adjacent peninsula was, and remains today, via a dirt track that branches off the main highway at the south end of the bay and clings to the east side of the bay. His map covers 450 square miles (1,165 km^2) of stupendous geology. Most of his fieldwork was conducted on foot. Provisions were conveyed to the field from a base camp on the peninsula by a three-burro pack train. Supplies and mail were delivered to the base camp by sailboat. Field mapping was conducted over a period of three months through the first part of the year in 1964.[2]

Before satellites with ground-positioning systems (GPS) interfaced with a workable digital elevation model (DEM) became available, the standard way to measure the height of a mountain was to use a barometer. But because ambient air pressure is subject to frequent change with the weather, a barometer must constantly be recalibrated with a known value such as sea level. McFall used a pocket barometer to take measurements at more than a score of localities spread through extremely rough terrain on the Concepción Peninsula. About halfway out the length of the peninsula, the highest peak reached by McFall was atop a huge igneous intrusion of tonalite, where he recorded an elevation 2,434 feet (742 m) above sea level. A topographic map issued by the Mexican government in 1980 identifies this part of Concepción Peninsula as the Sierra Gavilanes (Hawks Mountain), with a maximum elevation of 2,362 feet (720 m). In short, McFall demonstrated great initiative and physical stamina. He worked wonders with the tools at hand and produced a map that excited further attention.

During a January 1992 field course, Jorge and I brought a mixed dozen of American and Mexican students south from Santa Rosalia past Punta Chivato to make camp at El Requeson. The place features a tombolo located deep within Bahía Concepción that connects an islet of andesite with the bay's adjacent western shore. Shallow waters on opposite sides of the 1,132-foot (345-m) natural causeway flood across the center with each high tide. Marshall Hayes, the first of many thesis students to work with me in Baja California, conducted a detailed census on the intertidal faunas living on rocky shores around the circumference of Isla Requeson (Hayes et al. 1993). Thus, there was much to show the students. On the penultimate evening at El Requeson, Jorge produced a memorable *carne asada* dinner. Afterward, we began to discuss the possibility of new fieldwork across the bay on the Concepción Peninsula. That is when Jorge reached into his satchel and pulled out a copy of McFall's map and report.

Let no mistake be made. Our interests are skewed toward the arrangement of past shorelines on the Gulf of California and how they developed over time. Among other issues related to the kinds of sediments off former coasts is our interest in why the distribution of past marine communities thrived in some places but not others. It all amounts to a kind of obsession over the way life arranges itself in conjunction with the seascapes it occupies. Working on a local to regional scale, we are motivated to make geographical surveys that compare the past and present distribution of life in the shallow-water realms of the Gulf of California and its various islands. Other geologists may surpass our enthusiasm for the technical details of igneous and metamorphic rocks, but we are engrossed in how those rocks promote physical variations in a coastline.

Thus, when Jorge unfolded McFall's map under the glare of a propane lamp that evening, we immediately focused on the base of the Concepción Peninsula adjacent to the innermost, southeast corner of the bay. Covering about 23 square miles (60 km^2), the portion of the map we were most interested in barely amounts to 5 percent of the total region mapped by McFall. What we recognized was a former extension of Bahía Concepción expressed by crisp boundaries between younger sedimentary rocks of Pliocene age juxtaposed against various sorts of igneous rocks that formed earlier during the middle to late Miocene (figure 16).

Labeled on the map and briefly described in the report, the marine sediments that invaded the base of the Concepción Peninsula were said to include sand, marl, and coquina beds. From the map, it looked as if a chain of former igneous islands guarded the passage from the modern southeastern shore of Bahía Concepción to the inner bay. The map was drawn with such attention to detail that a narrow paleovalley cut in Miocene andesite for more than a half mile (almost 1 km) could be distinguished as filled with marine sediments (figure 16, arrow). Our minds were made up. The students would have a free day to swim and relax, while Jorge and I drove over to investigate the situation. We were already close by. To reach our destination, all we had to do was cross the lowlands at the south end of Bahía Concepción, a distance of about five miles (8 km) from the paved highway. How hard could it be?

As it turned out, it required several hours to reach the mouth of an arroyo that led east from the beach to the heart of Bahía Concepción's Pliocene strata. The original design of the 1971 highway

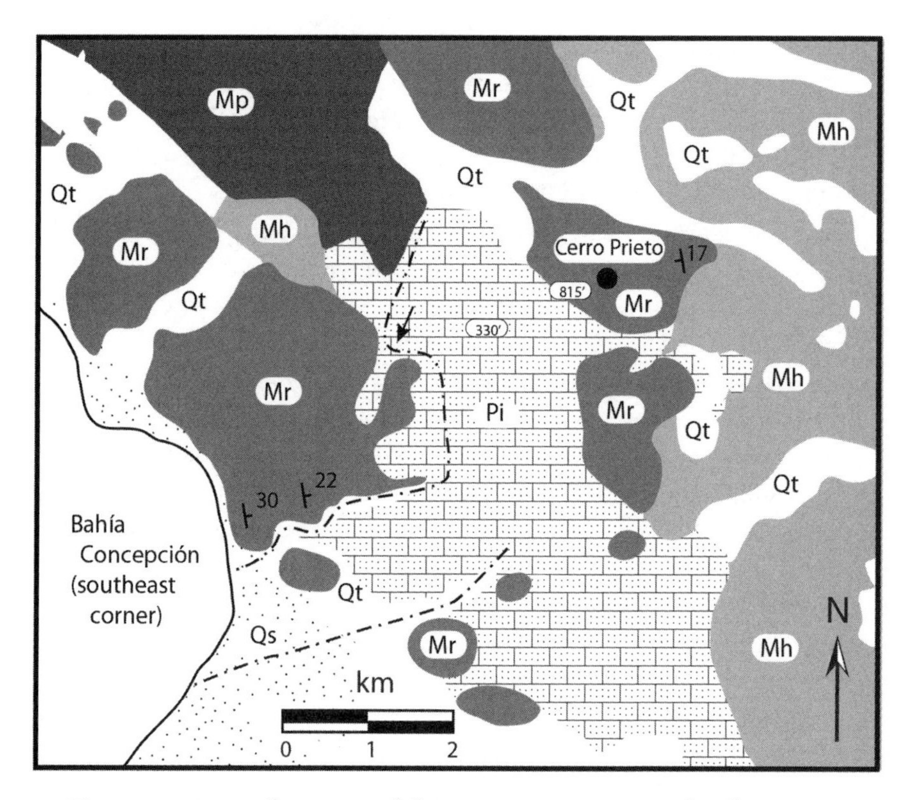

Figure 16. Revised portion of the reconnaissance map for the region around Bahía Concepción after C. Carew McFall (1968). *Note:* Arrow points to paleovalley; dashed-and-dotted lines indicate streambeds; and standard symbols for strike and dip enclose the area of McFall's "downwarp."

(Mexican Federal Highway 1) called for strategic stopping places along the way, where travelers with trailers or camping vans might be accommodated overnight. One such facility was built on the lowlands at the south end of Bahía Concepción. It was a poor choice. Not only is the place subject to episodic winter winds that funnel south down the long axis of the gulf and straight into the open mouth of Bahía Concepción, but those winds generate waves that pile up against the terminal shores of Bahía Concepción and spill onto the adjacent low ground. All that remains of the former trailer park are a few concrete pilings. Our initial attempt to follow the dirt track beyond the moribund park to the bay's opposite corner quickly brought us to a halt on muddy ground unable to bear the van's weight. Although the day was clear and calm, the winds and waves had visited not long before. We would become hopelessly mired, with the wheels sunk to the axles, if we continued by this route. Back on the highway, it took a few trials and errors to find the right connection to a gravel road over higher ground that would take us safely where we wished to go.

On reaching the far corner of the bay, we undertook additional scouting to identify the correct arroyo where we should continue inland on foot. The map indicated that we could expect two and a half miles (4 km) of hiking through a bend in the arroyo, before arriving at the mouth of the little paleovalley choked with marine limestone (see figure 16). The most abundant volcanic rock surrounding the inner marl and coquina beds is shown as the "Ricasón Formation," named for the same kind of andesite and basalt flows with coarse agglomerates that formed the islet near our encampment at El Requeson. McFall labeled outcrops exposing these rocks as "Mr" for the Miocene Ricasón Formation (a phonetic spelling of *requeson*, Spanish for cheese curd). The lumpy nature of a pyroclastic flow carrying rounded-to-angular blocks might merit comparison to cheese curds, but the place-name actually refers to the tombolo at El Requeson formed by the bleached-white remains of countless rhodoliths broken and reduced to coarse biogenic sand by wave action.[3]

Other details from McFall's map provided valuable insight. Strike and dip symbols affixed on the map for the Ricasón andesite, now immediately north of us on Cerro el Mono, showed individual flows dipped between 30° and 22° to the east (see figure 16). Such a relatively high angle in these layers would not have been in character for the original volcanic flows when they swept across the terrain about 10 million years ago. Some inclination assists flows to move with the help of gravity, but only the slightest slope is required. Likewise, a strike and dip symbol on the east flank of Cerro Prieto showed that the Ricasón andesite at that locality dipped 17° to the west. The dip angles recorded for these rocks told us that the layers were tilted in opposition to one another during a tectonic event long after the flows cooled. In effect, the "downwarp" that McFall (1968) cites in his report represents a fault-related structure imposed on the landscape.

As we hiked the entrenched arroyo at the mouth of the canyon with Bahía Concepción behind us (map 5), Cerro Prieto remained out of view. We expected to see it once we came to the prominent upstream bend ahead of us, and it would remain the central landmark in the day's excursion. So began an afternoon's march that brought unexpected revelations. We hadn't long to wait, because the earliest sign of limestone cliffs appeared a short distance inland banked against andesite on the north side of the arroyo. Pliocene fossils were abundant, the most prominent being the oversized conch, *Strombus galeatus*, which survives today throughout the gulf as one of the largest gastropods in the eastern Pacific Ocean.[4] The living conch prefers a subtidal, shallow-water environment with a sandy bottom. A few curio shops in La Paz stock shells of this large gastropod for sale (figure 17a, b). Lacking the dark brown outer covering of organic material (periostracum), the fossil relic of its kind sported larger ornamental nodes (figure 17c, d). Otherwise, the size and shape of the Pliocene representative are entirely consistent with those of its descendant. Its presence here as a fossil told us that something like a shoal between andesite islands marked the passage between the inner Pliocene bay and the present corner of Bahía Concepción.

As we continued westward along the streambed, here nearly 60 feet (18.5 m) wide, something caught my attention as fundamentally strange. One side of the arroyo was blanketed by normal gravel-size outwash consisting of mixed andesite and basalt pebbles, all dark in tone. In contrast, the other side was covered by bits of eroded, white chert (SiO_2). A sharp median line divided the two lateral halves of the streambed right down the center. The scenario bought to mind stories about separate waters from the confluence of the White and Blue Nile rivers in Africa. Waters come together from two different sources (clean versus muddy) and require some time before thoroughly mixing downstream. In fact, the division on the dry streambed was a clear signal that distinct materials were washed into the arroyo from different sources upstream.

The peculiar chert trail came to an abrupt end when we reached the bend in the arroyo. To our left, the streambed continued north and carried rounded pebbles of andesite and basalt almost exclusively. To our right, a gulley led to a notch-like feature cut into the rocks. Indeed, we later called this spot the "Notch," a place where erosion left a passage linking parts of two distinct subbasins within the landscape. Igneous rocks conforming to the Miocene Ricasón

Figure 17. Comparison between modern (*a, b*) and fossil (*c, d*) shells of the same gastropod species (*Strombus galeatus*). Photos by B. Gudveig Baarli.

Formation supported one shoulder of the declivity. This outcrop, however, was not marked on the McFall (1968) map. Most surprising of all, the gulley's opposite embankment was chert. The chert is massive, nearly 10 feet (3 m) thick, with little sign of individual beds. Above follow more chert layers, some fragmented and rather less consolidated than others that are more solid but brown in color. We were stupefied, because thick chert deposits are seldom encountered as laterally extensive rock formations. More commonly, limestone layers incorporate nodules of chert as a secondary by-product. Our amazement extended to the fact that the McFall (1968) map gave no advance warning about what was obviously a huge chert body.

Under certain circumstances, chert deposits may accumulate as the primary product of small, single-celled organisms with siliceous tests, called radiolarians. In this case, coastal upwelling of nutrient-rich water enhances the productivity of the radiolarians. Technically, such rocks go by the name *novaculite*. If this deposit were a novaculite, we would expect to find some evidence of siliceous microfossils. A preliminary look with a hand lens yielded no such sign. We would not be able to arrive at a definitive answer on the spot, but some chert samples were collected for further analysis.

Curiosity brought us back to the streambed for a short march north to the spot indicated on the map where a narrow valley was purportedly buried in limestone (see figure 16, arrow). Instead, we found the site chocked with rough conglomerate formed by large chunks of andesite held in place by a solid matrix of chert. It was more than enough for one afternoon. We hurriedly retraced our path back to camp at El Requeson to share the news of our discoveries with the students and announce a new research project for the following year.

As it happened, the project stretched over two field seasons: March 1993 and January 1994. The first was limited to a field party of four: Jorge and myself, each bringing a student assistant. Mine was Mark Mayall, a double major in geology and Spanish. Jorge's assistant was Sonia Gutierrez, a master's student. Sonia was from La Paz, and the four of us agreed to meet there on March 11, 1993. Jorge had contacts at the university in La Paz and assured me that we would procure a truck for the trip north. The vehicle turned out to be a small bus normally used to transfer students between La Paz and the university campus on the outskirts of town. We had plenty of room for our gear, and the bus came with its own driver, who slept in it at night. Getting the bus off Mexican Federal Highway 1 and over the rough roads to the mouth of the canyon at the southeast corner of Bahía Concepción was a challenge, but one our driver accepted with aplomb.

During the second excursion, we returned with a combined team of eleven students from our respective institutions. Our colleague

John Minch also joined us. We were determined to cover an area sufficient in size to map the chert body and better understand its geological relationships. Student squads, equipped with copies of topographic maps, were sent out in different directions to record what they found. The whole process involved the coordination of multiple day hikes that eventually resulted in the publication of a refined map and detailed report (Johnson et al. 1997). The trek laid out in map 5 with locality stops marked by stars is representative of a reconnaissance hike taken in 1993; it follows a loop from the Notch to the north side of Cerro Prieto and back. The outbound five-mile (8-km) hike follows parts of an abandoned road to San Sebastián on the east side of the Concepción Peninsula, while the return (a comparable distance) utilizes part of the New San Sebastián Road but also cuts back to the Notch cross-country.

FEATURE EVENT

Journey to the Far Side of Cerro Prieto

Although base camp is seven miles (11.5 km) east across the lower end of Bahía Concepción from busy Mexican Federal Highway 1, the night is punctuated by a succession of tractor-trailers with loud mufflers that issue a staccato burst when the driver eases the throttle on the winding coastal road. The noise carries easily across the water, enough to jar the dead from the heaviest of slumber. I long for the day to begin. There is an edge to the morning's duties, as if they cannot be discharged fast enough to begin the day's essential task. Soon enough, however, we are assembled and enter the canyon flanked by Cerro el Mono (Monkey Hill) to trudge through deep gravel. I find better traction for my footing near the median line in the arroyo, where broken chert fragments spill against more-rounded pebbles of andesite and basalt. With the Notch as our first objective, we stream past the limestone outcrop that forms part of the arroyo's north border. We already know what fossils the limestone holds and what they signify.

A tangle of thick brush blocks the opening to the gulley at the bend in the arroyo. Pushing through to the world hidden on the other side beneath the Notch requires determination. But we have

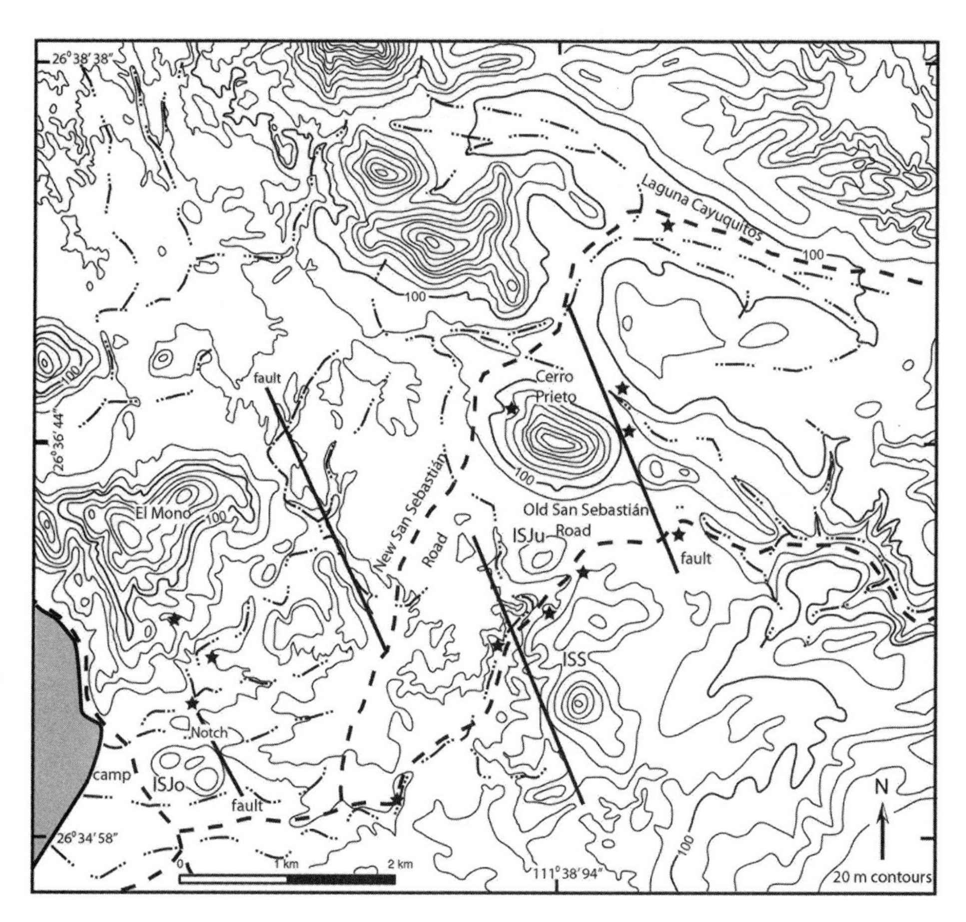

Map 5. Topography and geology at the base of the Concepción Peninsula. *Note:* Old and New San Sebastián roads are marked by heavy dashed lines; dashed-and-dotted lines represent streambeds; and black stars denote key stopping points along the trek. *Paleoislands:* ISJo = Isla San Jorge; ISS = Isla San Sebastián; ISJu = Isla San Juan. Map by author.

been here before, and care is taken to bring everyone through the snarled vegetation, one by one. The scene beyond opens dramatically to a spacious amphitheater eroded in crumbly layers above the massive foundation of the chert formation directly east of the Notch (map 5; figure 18). It is early, but the bright sunshine catches a multitude of facets in the crumbly chert to create a radiant surface of blinding white. It is painful to the eyes, and the lingering spell of

Figure 18. Amphitheater eroded in crumbly layers of a chert formation; Jorge Ledesma (lower right) for scale. Photo by author.

a sleepless night presses an added heaviness. We spread out along the amphitheater, each to seek a different level, where the meaning of the chert might be uncovered.

I am drawn to a singular brown layer with burrow-like traces. They remind me of the embayment floored by carbonate mud behind the reef fronting Mary Creek on St. John in the US Virgin Islands. It is as if I stand in waist-deep, turquoise water with my ankles sunk in the soft mud, shielding my eyes from the sparkling reflection of the sun off the water's surface. Using a facemask, I was able to glide over the lagoon bottom and examine small, volcano-shaped mounds of excavated sediment that mark the entrance to burrows made by the Caribbean ghost shrimp (*Callianassa* sp.). My impulse to find rational comparisons between the present world and the geological puzzles that confront us is a habit. If the structures before me are the fossil burrows they appear to be, then an appropriate scientific name for the trace-fossil taxon might be *Ophiomorpha*. Literally, the name translates as "snake-shape," but the words relate to winding, horizontal cavities below the vertical entrance to a burrow system. Here, the preservation of fossil burrows in the brown chert reminds me of the vertical tunnels made by the ghost shrimp. In July 1992, Mark Mayall and I spent several days mapping *Callianassa* burrows on Mary Creek, where we found a density of up to three or four entrances per 10.75 square feet (1 m^2) in shallow water more than 5 feet (1.5 m) deep.[5]

Jorge crouches in the shade near the Notch, using a pickax to dig into the outcrop. "Over here," he yells. "I've got something." We crowd around and kneel on the debris Jorge scraped from the cliff. The clean face of a chert layer reveals a horizontal line of fossilized wood extending nearly a couple feet (60 cm) in length. Some broken pieces have fallen from the outcrop onto the talus. They look like so many stem fragments from a sixteenth-century tobacco pipe recovered from an archaeological dig: cylindrical bits that are hollow inside. As petrified wood, the bits have been silicified. They are darker than the chert in which they were entombed, and their surface reveals a grainy texture with small scars that look like knots. When held under a magnifying glass examined in cross section, cell structure can be recognized around the inner rim. The pores resemble xylem tubes that carry tree sap. However, these woody stems were not buried vertically, and they do not branch.

The petrified wood is not from the aboveground part of a tree, but rather, its roots. Entirely horizontal in alignment, these make a good match with the subsurface runners belonging to the black mangrove (*Avicennia germinans*). Nearby at El Requeson, the habitats of black and red mangroves (*Rhizophora mangle*) can be compared side by side. The air-breathing pneumatophores of the black mangrove that rise from horizontal runners buried in the sand between adjacent trees are easily distinguished from the thicker prop roots of the red mangrove exposed slightly seaward at low tide. True, we find no evidence of the distinctive pneumatophores in the chert deposit, but the size of the fossil runners gives pause for thought. A combination of shallow-water burrows and black mangrove fossils in different layers of the chert deposit begins to bring the setting to life as a former lagoon.

It remains to be seen just how large an area the exposed chert deposit occupies beyond the natural amphitheater, which is capped by limestone to the east. Turning southeast, we cross through the Notch and explore more of the chert deposit for three-quarters of a mile (1.2 km) until reaching the dirt road to San Sebastián. No other fossils are encountered, but the same limestone cap rock that fringes the chert body at the amphitheater appears in thin beds along the road. Basal layers of the limestone contain pebbles of chert eroded from the underlying chert body. Continuing another third of a mile (500 m), we follow the road to where it starts to climb the limestone, but we leave the road to trace the limestone around a bend to the east.

The limestone is massive, lacking obvious fossils. Above, there appears red mudstone, 3.28 feet (1 m) thick, crowded with root casts (figure 19). These are much larger than the horizontal runners found in the chert beds below, and the shapes generally are divergent downward. Closer examination of a cross section through one of the structures (figure 19, inset) shows a pattern of circular growth rings. Another limestone formation follows above the mudstone, almost as thick and massive as that immediately below the mudstone. Clearly, the mudstone with its extensive root casts represents the interruption of a marine incursion signified by the limestone. For sake of simplicity, we agree to call these two units the lower and upper limestone beds. A mass of roots much like that of the red mangrove marks the natural interlude between two phases of limestone deposition in a secluded lagoon off the main part of Bahía Concepción during the later Pliocene.

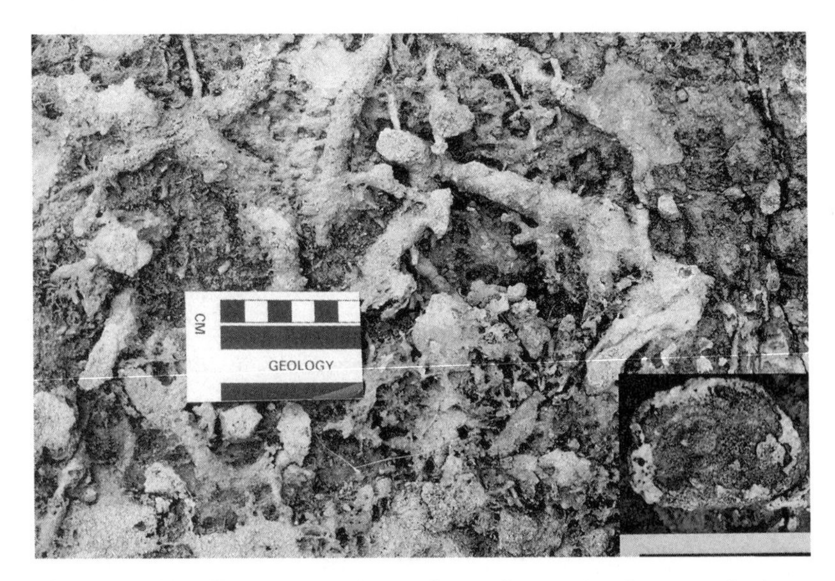

Figure 19. Fossil root casts. Inset photo shows circular growth rings detailed in cross section (scale in centimeters). Photo by author.

This tells us nothing more about the chert body buried below but does provide a sense of spatial limitation to the marine waters that flooded the base of the Concepción Peninsula. The McFall map (figure 16) indicates that limestone can be traced at least three and three-quarter miles (6 km) east across the peninsula from a connection with contemporary Bahía Concepción. Given that we now stand well above the water surface of Bahía Concepción, what is the likelihood that a rise in Pliocene sea level coincidental with local faulting may have penetrated along previously existing valleys across the peninsula from both the west and the east to form an island? This is an intriguing thought that gives added impetus to the day's adventure.

The old road to San Sebastián, which was made impassable by one of the tropical hurricanes that strike the Gulf of California every few years, runs parallel to a major arroyo. Taking advantage of the old road, we enter a deep canyon and make our way northward for nearly a mile (1.5 km). Canyon walls on our left rise up more than 170 feet (52 m) and include many of the rock units with which we have become familiar (figure 20). A relationship new to us is a prominent, 88.5-foot (27-m) conglomerate composed of angular, pebble-to-boulder-size clasts of andesite cemented within a matrix of red siltstone. Leaving the road to descend to the floor of the canyon, we find that the conglomerate makes an unconformity on the igneous Pelones Formation of McFall (1968), characterized by him as andesite agglomerate. One of the students photographs Jorge as he stands on the unconformity surface (see figure 20). The McFall map (figure 16) shows a large exposure of the Pelones Formation to the northwest (marked by the abbreviation Mp), but here it is almost completely covered by the overlying conglomerate, as well as succeeding lower and upper limestone beds separated, again, by thick, red mudstone. According to McFall's (1968) scheme, the igneous Pelones Formation is older than the Ricasón Formation. Thus, we assume that it extends eastward beneath the hills above the road to our right.

The washed-out remains of the old road climb steeply above the arroyo floor, bringing us northeast. I spy a ledge of limestone against the shoulder of the andesite hill above us to the right, and I leave the road to investigate. The limestone, which we equate with the lower limestone unit, contains abundant fossil oysters. Some are weathered free of the limestone and lie loose on the surface. They are the largest

Figure 20. High canyon walls on the Old San Sebastián Road (Jorge Ledesma and student for scale). Photo by author.

I have ever seen, 14 inches (44 cm) in length. Shucking a living oyster this size would provide quite a meal. Back on track, we continue upward to where the road levels off. Cerro Prieto is due north of us, its dark summit rising 590 feet (180 m) above our position. Closer at hand on both sides of the road are matching limestone outcrops pressed against opposing hillocks of andesite. The oyster limestone above the road, behind us, is at the same level as the limestone on our left. The junction where the limestone abuts the andesite rocks of the Ricasón Formation marks a former shoreline. Andesite boulders fixed in the limestone at this locality have attached fossil barnacles. To the west, a similar tan limestone is exposed against the flanks of the opposite hill. Hillocks formed by igneous rocks protrude above the limestone, and the realization dawns that the elevations denote small islands. Cerro Prieto was probably a much larger, higher island.

The walking now is easier on the flats. We follow the road farther northeast for a half mile (0.8 km), to where an isolated knob of limestone rises from a position south of the road. It is an outlier, an erosional remnant separated from comparable rocks. The limestone is massive, more than 15 feet (4.75 m) thick, and contains both fossil oysters and abundant pectens. Here, the flooded landscape was in more of an open setting away from surrounding islands. Looking back to the southwest from our perch atop the outlier, we can readily see how limestone bands wrap around two sides of a former island. Eastward, the Old San Sebastián Road begins its descent to the outer coast of the Concepción Peninsula. Checking our map lets us see that we are almost exactly midway between Bahía Concepción and the open Gulf of California on an east-west axis. To continue eastward in search of more limestone is tempting, but Cerro Prieto beckons from the north. To what extent is it surrounded by limestone?

A faint track turns northwest off the Old San Sebastián Road, passing a low saddle between Cerro Prieto and an adjoining andesite hill to its east. We choose this less-traveled path and strike across level country. There is no limestone underfoot, and soon we reach the 262-foot (80-m) contour (see map 5), where the track begins to rise onto the saddle. Parts of the track are worn to bedrock. According to McFall's map (figure 16), andesite of the Ricasón Formation is exposed here. The first sedimentary rocks encountered below the saddle on the northeast flank of Cerro Prieto are bizarre looking. They are a red mudstone in thin layers penetrated by columns with crumbled margins. Reminded of contemporary hot springs in the shallows of Bahía Concepción, we jokingly refer to these rocks as "champagne mudstone," disturbed by rising gas bubbles from a Pliocene geothermal spring.

Such springs commonly are associated with fractures, and we speculate that a fault is responsible for the worn saddle in the landscape just traversed. Such a fault might account for the isolation of the limestone outlier explored on the south side of the Old San Sebastián Road. A line from the margin of the outlier through the saddle to the "champagne mudstone" (see map 5) follows a bearing (N 22° W) close in alignment with one of the master faults demarcating the east shore of Bahía Concepción. Now we have another idea to work with, one that may help to explain the chert body as a deposit altered by silica-rich thermal waters leaked from a parallel fault.

Now below the crown of Cerro Prieto, we gaze northward across a trough-shaped depression eroded by two arroyos that diverge in opposite directions. One drainage system carries water from ephemeral rainfall to the west, ultimately to flow into Bahía Concepción. The other leads eastward, to empty into the Gulf of California at San Sebastián. In effect, we stand astride Cerro Prieto poised at a peninsular divide. The east drainage cuts through a layer of limestone, clearly visible from afar. Just as clearly, the McFall map (figure 16) fails to record the limestone at this location. With renewed enthusiasm, we set out to cross the trough and reach the limestone outcrop about a fifth of a mile (300 m) away. The outcrop is crowded with fossil oysters. None are as large as the individual shell found farther south, but these were prolific. A few steps northward beyond the embankment outcrop, the limestone abuts andesite rocks. Only a single bounding step is needed to cross from the Pliocene limestone to the older, Miocene andesite of McFall's Ricasón Formation.

The trough-shaped depression was once a narrow lagoon flooded with seawater, perhaps brackish but capable of supporting an enormous population of oysters. The oysters thrived in close proximity to sea cliffs formed by andesite rocks. The same limestone must have occupied the entire depression flush with Cerro Prieto, but the limestone embankment on the south side of the arroyo was removed long ago by erosion. The bounce in my steps that sent me across the geological contact, vivid in beige limestone against red igneous rocks, catapults me several steps beyond. Unexpectedly, something appears before me that I had previously only read about. The shock sets the smallest hairs on the back of my neck on end.

My next step brings me to andesite boulders formed in a perfect ring approximately six feet (1.83 m) in diameter. Aboriginal inhabitants of this region knew this spot and had camped here. Was this simply an overnight bivouac, or was it a magic circle where an individual came to fast and dream? Either way, the earliest explorers of the region were familiar with the layout of the landscape where the drainage etched a natural pathway across the peninsula from one side to the other. I am no archaeologist, but the physical connection with an artifact of a vanished people sets my mind afire.

The principle of cultural relativism tells me that I should resist the urge to draw uninformed parallels between my comprehension of nature and that of another people, past or present. It is a temptation hard to resist. Heaps of shells and other debris left behind by nomadic tribes up and down the gulf coast testify to the bounty of the sea as a renewable food source. No doubt, aboriginal inhabitants were familiar with the dominant species of local shellfish. The waters hereabout certainly favored rock oysters and related species. How would a "primitive" mind interpret a fossil oyster? The loose shell eroded from a rock formation in which it was once buried might readily be regarded as a piece of trash carried inland by a foolish person (especially if it was a large and relatively heavy shell). How about a whole trove of oyster shells, including valves still united, held together in a solid mass exposed by a stream far from the seashore? The ring bears witness to the fact that other humans paused here long ago. Was this particular stopping place one only of convenience, being located halfway between two bodies of saltwater? I am struck by the fact that the ring is sited close to the colorful junction between limestone and andesite. My imagination wants me to find kinship with an individual from the past who understood that the sea, too, once paid a visit to this lonely place.

The day already has slipped past the noon hour, but we are unable to take our lunch break anywhere near the ring. A feeling that we intrude on a private space presses strongly on us. A small measure of shade might be found in the arroyo at the foot of the oyster limestone. We do not retrace our steps. Instead, we press onward to cross the divide between the two drainage systems. After a little more than a half mile (0.8 km), we stumble onto the New San Sebastián Road, where it enters a northeast-directed valley and continues inland along the arroyo of the Cayuquitos drainage (see map 5). There is not a shred of shade anywhere. Lunch is quickly consumed. It is scarcely more than a break to slake our thirst.

The McFall map (figure 16) concurs with our topographic map (map 5) to show the junction with another east-west valley ahead. In this case, however, two branches of the arroyo converge from opposite directions. Contrary to the equally narrow valley behind us, the more northerly valley is not bifurcated by the peninsular divide. It is fully enclosed within a single catchment basin. We do not have far to go, less than a mile (1.5 km) around a bend in the road. The limestone is uncharted on the McFall map. As found before, it includes fossil oysters. This appears to be the most remote of the limestone lagoons hidden within the landscape of the Concepción Peninsula. We call it Laguna Cayuquitos. The road continues eastward across the valley for another one and a quarter miles (2 km), before crossing the 330-foot (100-m) contour and rising toward the peninsular divide. The limestone conforms to the shape of the valley bottom as far eastward as the eye can trace. There will be other exploratory excursions in the days ahead, but the moment has arrived to turn back for camp.

The return trip, on the New San Sebastián Road, skirts the west side of Cerro Prieto. A brief stop to explore a prominent dent in the flank of the great hill is a strong enticement. If nothing else, the opening in the side of the hill may provide some needed shade. According to the McFall map (figure 16), the entire edifice is formed by andesite belonging to the Ricasón Formation. When it is viewed from the road, however, the sun illuminates what appear to be stratified rocks far below its towering 788-foot (240-m) peak.

Near the road, we encounter the slope and begin a short but arduous climb. After about 82 feet (25 m) of ascent, our goal is reached, and we learn that the layers spied from below are indeed stratified rocks plastered against the side of the hill. The discovery reminds us how the day began at the Notch, where immediately above the massive base of the chert formation we investigated a succession of crumbly chert layers. The layers are not as resplendent here and are more beige in tone. There is no mistaking the siliceous content of these beds, which amount to a thickness of 6.5 feet (2 m).

Our prospect from the flank of Cerro Prieto is sufficient to scope the pale blue waters of Bahía Concepción in the distance. We also see the white gash where the natural amphitheater near the Notch exposes the big chert body. The layers on which we now stand are the remains of a distinctly younger chert halo that once surrounded Cerro Prieto at this level. Indeed, the layers trail off to the north and appear to wrap around the north face of the hill. Putting aside the issue of secondary replacement of former carbonate rocks by silica-rich thermal waters, these layers imply submergence of the surrounding landscape to a degree not previously appreciated. Was the relative rise in sea level enough to sunder the Concepción Peninsula from the rest of the Baja California peninsula? To prove the point, correlative layers of the upper chert formation would need to be located at about the same elevation on surrounding hills such as Cerro el Llano to the north and Cerro el Mono to the west.

Plate 1. Granite boulders and desert plants, including boojum trees, cardón, and cholla, southeast of Cataviña. The body of water in the far distance is Laguna Chapala. Photo by author.

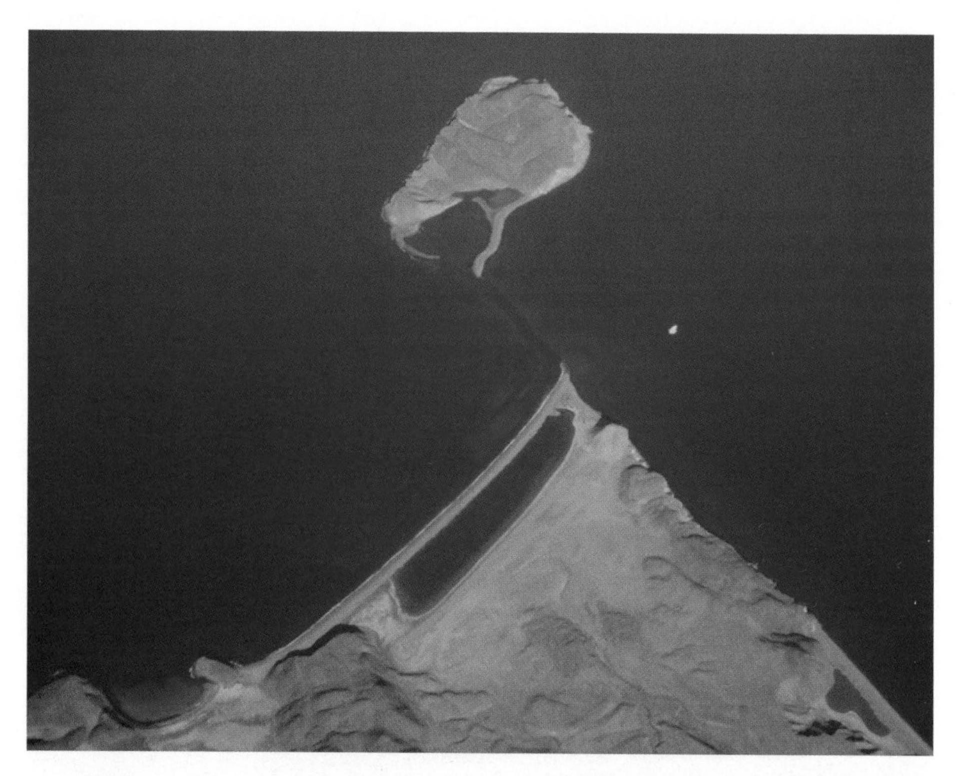

Plate 2. View from 30,000 feet (9,144 m) to the northeast over the south end of Isla Angel de la Guarda featuring two closed lagoons. Photo by David H. Backus.

Plate 3. View north from near the top of the great sand ramp with Bahía de los Animas and Isla Angel de la Guarda in the distance. Photo by author.

Plate 4. View east from 30,000 feet (9,144 m) over the great sand ramp with climbing dune to the left and falling dune to the right (marked by white arrow). Photo by author.

Plate 5. View east from 30,000 feet (9,144 m) over the San Francisquito basin with Ensenada las Palomas to the left (north). Photo by author.

Plate 6. View north from the San Francisquito road over the exhumed unconformity between Pliocene limestone and granite basement rocks. Cardón cacti (center) grow from thin soil on the granite. Photo by author.

Plate 7. View east from 25,000 feet (7,620 m) over the south end of the Concepción peninsula and its lost Pliocene lagoons (beige tones surrounded by red). Photo by author.

Plate 8. View southwest from Punta el Mangle over the wedge of Pliocene limestone (white rim) abutting the Cerro Mencenares volcanic complex (red). Photo by author.

Plate 9. View north to Punta el Mangle showing red discoloration on El Coloradito Fault. Photo by author.

Plate 10. View east from 20,000 feet (6,096 m) over the radial topography of the Cerro Mencenares volcanic complex. Photo by author.

Plate 11. Humboldt squid (*Dosidicus gigas*) in death throes (approximately 3 ft. or 91 cm long). Photo by author.

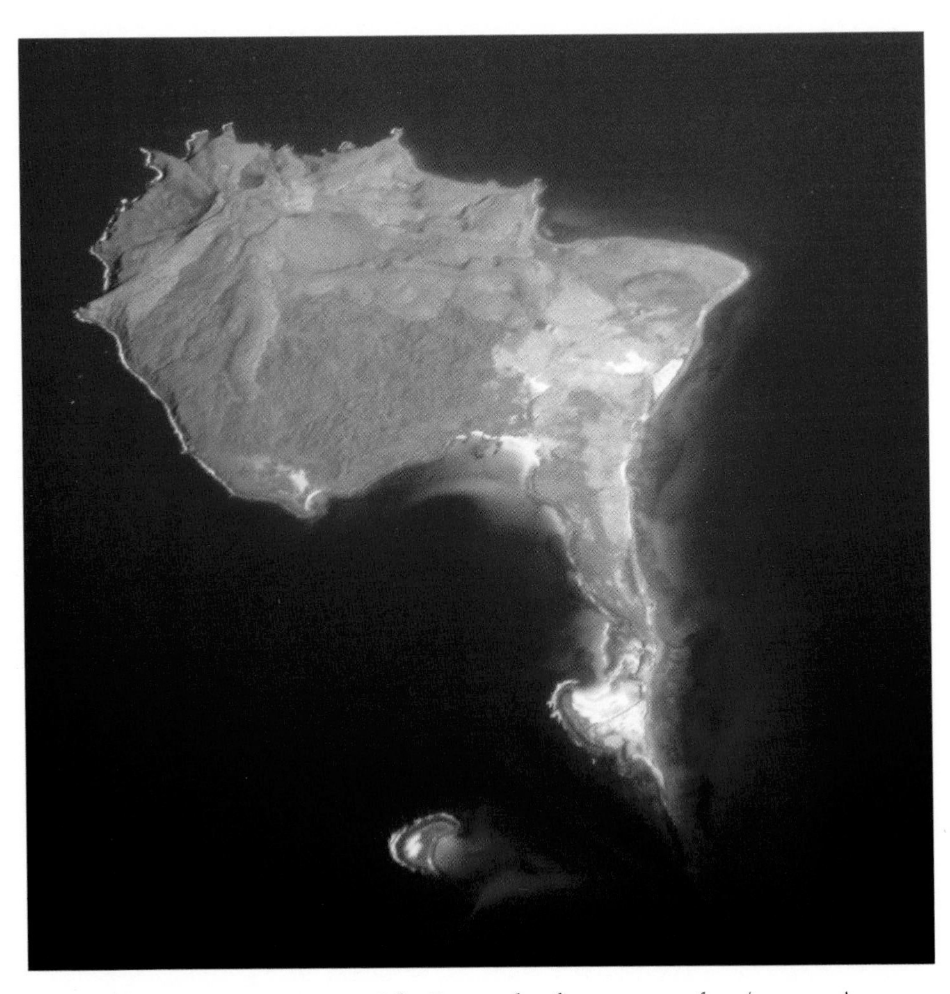

Plate 12. View east over Isla Coronados from 15,000 feet (4,572 m). Photo by author.

Plate 13. View north to a plateau 635 feet (194 m) high on Isla Monserrat with Pliocene strata seated on an unconformity over tilted Miocene andesite. Photo by author.

Plate 14. View southeast over Isla Monserrat from 30,000 feet (9,144 m) showing patches of limestone (white) on older volcanic rocks (red). Photo by author.

Plate 15. Paredones Blancos (White Cliffs) on the west coast of Isla Cerralvo. Photo by author.

Plate 16. Granite platform and overlying Pliocene ramp south of Los Carillos on Isla Cerralvo. Photo by author.

The long road south from Cerro Prieto crosses a limestone plateau dropping less than 135 feet (41 m) over the next two miles (3.22 km). Along the way, we traverse badly weathered limestone that belongs to the upper limestone unit. The western part of the plateau is cut by an arroyo trending N 22° W (see map 5), suspiciously parallel to the fault trend interpreted on the east side of Cerro Prieto earlier in the day. Passing the optimum spot where the arroyo's alignment can be measured, we make a hard right turn and hike due west across the landscape.

In eight-tenths of a mile (1.3 km), we reach the western edge of the plateau and start to ease our way down the escarpment. Mark Mayall loses his footing at the top and dislodges a large piece of limestone that clatters down the slope into the streambed below. Taking care, we measure the rock layers through which we descend. The sequence amounts to nearly 110 feet (33.25 m). At the top is upper limestone (14.75 ft. or 4.5 m thick). It is underlain by red mudstone (13 ft. or 4 m thick), which in turn is underlain by the lower limestone (8.25 ft. or 2.5 m thick). More mudstone (8.25 ft. or 2.5 m thick) separates the overlying limestone from white chert beds (27 ft. or 8.25 m thick) beneath. At the bottom of the sequence is 38 feet (11.5 m) of conglomerate composed of andesite pebbles and cobbles. The overall succession is not as thick as that encountered earlier in the day, where the Old San Sebastián Road passed through the canyon (see figure 20), but it is the same in sequence.

Hoping to find some fossils, I stride over to the limestone block dislodged by Mark from the top of the escarpment. Turning the block over, I am startled to find a rattlesnake curled into a cavity of the limestone. It was hibernating within the recesses of the limestone. It is now dead. We agree to name the section locality, now measured, Rattlesnake Ridge (at 26°235'70" N, 111°40'5" W).[6] The arroyo in which we find ourselves is the same drainage stretching north-south between the Notch and the paleovalley originally mapped by McFall (figure 16, arrow). All that remains of the day's journey beyond Cerro Prieto is to complete the circle and return to camp by way of the canyon that initially brought us inland from Bahía Concepción.

In the follow-up to the 1993 and 1994 excursions, much was accomplished to answer key questions about sedimentary formations deposited at the base of the peninsula. Laboratory tests, for example, determined that two different silica sources enriched the huge, layered chert deposit in the stratigraphic succession. One was from airborne tuff that locally blanketed the area during volcanic explosions; the other came from saturation of silica-rich water via thermal vents that later converted much of the volcanic glass and all of the original carbonate sediments to opal (Ledesma-Vázquez et al. 1997). Notably, tests also revealed halite (salt: NaCl) as one of the mineral constituents in the chert. This helped confirm that the deposit had a marine link. Offset of beds at the Notch also told that the immediate area was cut by a fault oriented much the same as others in the region (see map 5).

Additional exploration, which included excursions farther north and east toward San Sebastián, allowed us to reconstruct how the region's paleogeography changed with the addition of different sedimentary rock units. This was done by plotting the extent of key units on a succession of maps (like frames in a film clip) to show the gradual expansion of marine flooding that reached all the way to Laguna Cayuquitos (Johnson et al. 1997). No evidence was found that either the lower or the upper limestone units ever breached the Concepción Peninsula during higher stands in Pliocene sea level. A prominent gradient in the distribution of marine fossils from the lower limestone shows that biological diversity was greatest closer to the connection with modern Bahía Concepción, and lowest (oysters only) farther inland. This served to reinforce the interpretation of physically remote lagoons with restricted marine circulation. In terms of physical geography, the most salient aspect was confirmation of former islands hidden in the landscape. Like explorers of old, we reserved the right to name newly discovered places. Thus, we added to the lexicon of names Isla San Sebastián (one of the larger islands) and Isla San Juan (one of the smaller islands), visible along the Old San Sebastián Road, and also Isla San Jorge near the Notch (see map 5: topographic elevations marked as ISS, ISJu, and ISJo, respectively).

While driving Mexican Federal Highway 1 along the stretch of road at the south end of the great bay, one can clearly see the lower limestone of the Concepción Peninsula in the distance for a few fleeting moments. It is easy to miss, but the flat-lying limestone layers leap out from the otherwise uneven landscape dominated by igneous rocks. More impressive is the view from commercial flights heading

Figure 21. Arrowhead fashioned from white chert derived from the massive chert deposit at the base of the Concepción Peninsula. Photo by Carlos M. da Silva.

south to Loreto (see plate 7). When the flight path is right, the lost lagoons of Bahía Concepción are easy to pick out by the pale reflections of the chert and limestone deposits. It is a remarkable landscape with unique stories that remain yet untold. During another visit to the Notch in 2010, one of our party stooped to pick up an arrowhead fashioned from the white chert (figure 21) brought to the arroyo by the most recent outwash. In an instant, I had the answer to a question that had long bothered me. Did the early human dwellers in this landscape know and appreciate the geologically distinctive white chert deposit as a valuable resource? The proof was indisputable.

6

Intersection of Fractures at El Mangle

If there is a defining moment in the formation of the peninsula . . . , it occurred 3.5 million years ago when a shift in the fissure between plates passed Baja California like contraband from the North American plate to the Pacific plate.

—*Bruce Berger,* Oasis of Stone

SEASIDE CLIFFS ON THE Gulf of California at El Mangle, some 18.5 miles (30 km) north of Loreto, command a fine view that inspired substantial investments in preparation for a grand hotel. Quite simply, it was a dream location too remote for fulfillment. Doomed by a long access road subject to episodic destruction by tropical storms, the buildings remained incomplete and lapsed into ruins over the years since 1998, when we first stumbled onto the place. Our excitement over the geological setting at El Mangle eventually led to three field seasons devoted to the area: January 1999, January 2001, and January 2002. It is normal practice to spend the last day or so during any given field project to scout for localities that might lead to new or expanded studies. After completing fieldwork in the San Nicolás area, we planned a final exploratory day at the conclusion of the January 1998 field season.

As a matter of fact, investigation of the San Nicolás region was an opportunity seized by Jorge Ledesma as an extension of our studies on the lower part of the Concepción Peninsula. Home to don Chico and his family (see chapter 1), the small village of San Nicolás is only five miles (9 km) south-southeast of San Sebastián (see map 1, locality 6). The territory around San Nicolás features another Pliocene basin, its geology left uncharted on the margins of the McFall (1968) reconnaissance map. I was exhausted after intense fieldwork south of San Nicolás at Punta San Antonio (Johnson and Ledesma-Vázquez 1999), followed by an exhilarating boat trip north from San Nicolás to the tip of the Concepción Peninsula. Fatigue beyond normal levels of endurance was exacerbated by don Chico's star rooster, whose instinct to compete with neighboring roosters induced hearty calls starting promptly at 2 a.m. every morning. A good dinner (revenge with *pollo con mole*) and a comfortable hotel bed in Loreto promised to be restorative after several sleepless nights at don Chico's family compound.

For many years now, a checkpoint manned by the Mexican army north of Loreto has obscured the turnoff from Mexican Federal Highway 1 leading east to the fish camp at San Bruno (map 6). No such military installation existed on January 21, 1998, the morning Jorge recommended we drive out from Loreto to San Bruno to hunt for potential Pliocene formations in the North Loreto basin. Because of a study by McLean (1989), we were already familiar with Pliocene strata in the South Loreto basin. Other colleagues, Becky Dorsey and Paul Umhoefer (both from Northern Arizona University at the time), were adding to McLean's knowledge of the South Loreto basin. Thus, we wished to explore new ground relatively untrammeled by geologists, a little farther to the north.

San Bruno of the Loreto region is a place shrouded in historical significance regarding the founding of missions by religious orders in the Spanish Californias. Thirteen years before the mission at Loreto was successfully established and became the de facto capital of the Californias, a failed attempt at settlement lasting little more than a year was made at San Bruno in 1684.[1] Eyewitness accounts of native inhabitants recorded by the expedition's military commander, Admiral Isidro de Atondo y Antillón, provide details about the dress, demeanor, and practices of a now-extinct people. The mission site had to be abandoned because of a lack of reliable drinking water. Nothing remains at San Bruno to suggest the intensity of initial contact between 55 visitors from mainland Mexico (many Europeans)

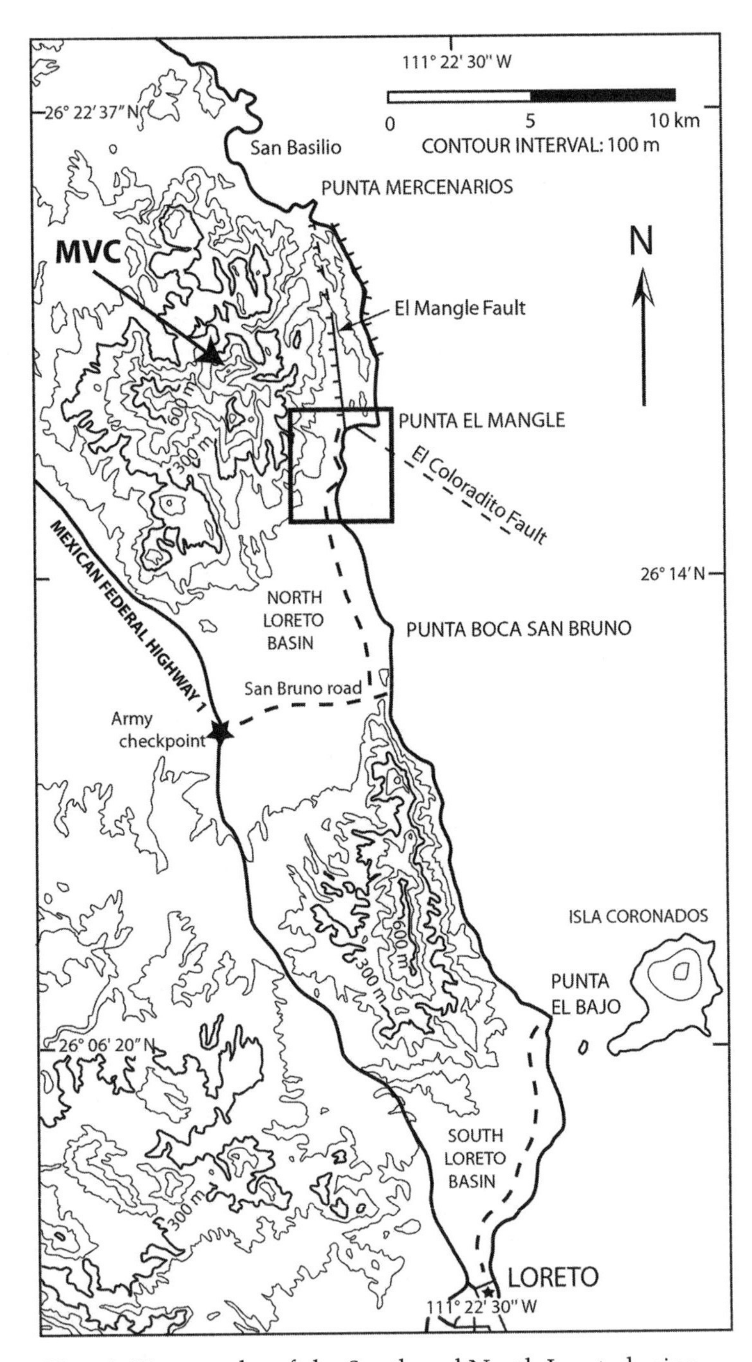

Map 6. Topography of the South and North Loreto basins related to the Cerro Mencenares volcanic complex. *Note:* Heavy solid line and dashed lines represent primary and secondary roads; black star marks the location of a checkpoint; black frame at Punta el Mangle marks limits of map 7. *Abbreviation:* MVC = Mencenares volcanic complex. Map by author.

and the region's approximately 2,500 Indian inhabitants.[2] Moreover, the geological nature of the area immediately surrounding the boat launch at the end of the San Bruno road is unexceptional.

The dirt road to San Bruno ends at the shore, passing between two hills with elevations that rise over 197 feet (60 m) above sea level (see map 6). Hence, scenic views north and south along the coast at that spot were limited. I was singularly unimpressed. Dog-tired and ready to call it quits for the remainder of the trip, I crawled back into the front seat of the van Jorge had brought from Ensenada. A southerly breeze was blowing, very unusual for that time of the year. It was good to come in out of the wind, sit still, and close my eyes.

Meanwhile, Jorge decided to answer the call of nature and chose to find privacy around the side of the more northerly hill. He was gone too long for that kind of business, and I began to be impatient. I was comfortable enough, however, and relished the chance to rest. When Jorge returned to the van, it was not to the driver's door but to the passenger's side. He opened my door, and the fresh salt air immediately filled the space. "You better come with me," he announced. "I think you'll like the view." The southerly breeze that day had conspired to set our research agenda for the next few years, because Jorge's line of sight (if not his aim) had been trained to the north in the direction of Punta Boca San Bruno (only 1.25 mi. or 2 km away) and toward Punta el Mangle on the horizon (see map 6). The unencumbered view along the shoreline took in one of the most magnificent limestone ramps embanked against older volcanic rocks that I had seen since our earlier studies at Punta Chivato. A jolt of enthusiasm instantly overcame my lethargy, and I agreed with Jorge that we had to find a road leading northward toward El Mangle in all possible haste.

As we drove on a lesser track parallel to the shoreline, every passing second brought us closer to the southeast flank of the Cerro Mencenares volcanic complex. In its entirety, this semicircular structure covers an area of about 58 square miles (150 km^2) and reaches a maximum elevation of 2,592 feet (790 m) above sea level (see map 6). Later, we learned that a group of Italian geologists had been busy studying the igneous petrology and structure of the enormous stratovolcano (Zanchi 1994; Bigioggero et al. 1995). Uplifted and exposed along the shores of the gulf, beds of limestone form a ramp inclined seaward to conceal the lower margin of the complex like an apron. The colors toward which we raced appeared suspended above the road like bold strokes painted on a vast canvas: beige to white bands molded against a reddish base, all framed by darker and lighter tones of blue from the sea and sky. Here were the raw remains of a former sea impounded against the rough shores of a long-quiet land from the past. As Dave Backus later put it when we returned in January 1999, the scene amounted to a full-up confrontation of "rocks in your face."

Where the road first deviated to meet the coast on our northward outing (maps 6 and 7), Jorge stopped the van, and we clambered out to examine sea cliffs eroded flush against the edge of Cerro Mencenares. The wind had abated, and the surface of the sea was dead calm. Tumbled blocks of limestone across the cliff face made the boundary indistinct, but the scene evoked a palpable sensation of riding ashore on a plank of limestone that thins to a feather's edge against the volcanic coast. Behind us to the south, the modern sea cliff was all limestone, and it reflected white in the bright sunshine. To the north, the andesite rocks of the volcanic complex absorbed sunshine on a dull-red surface. Like eager arrivals in a new land, we did not pause for long but returned to the road in full anticipation that it would bring us to better places.

One of those portentous gates that announces your arrival on the outskirts of a grand estate appeared around a curve in the road. Beyond the gates, there unfolded an extraordinary venue with a much-expanded limestone shelf neatly joined against the abrupt slopes of the Cerro Mencenares complex. In effect, the scene opened onto an entire "fossil" bay folded into a hollow on the flanks of the former volcano. To gain access, however, we would have to pass through closed gates: "Heaven's Gate," as we soon came to remark. A lack of space prevented us from driving around the barrier imposed by the heavy gates. A thick chain wended its way between and around stout bars. If the chain was padlocked, our journey was finished.

On closer inspection, we discovered that the chain could be unraveled and gates swung open to admit our vehicle. Ahead, the road hugged the margin of the volcano, and the limestone shelf grew ever wider, such that the waves lapping against the base of the cliffs disappeared from sight. For the next four-tenths of a mile (640 m), the shelf was covered by native vegetation, but thereafter someone

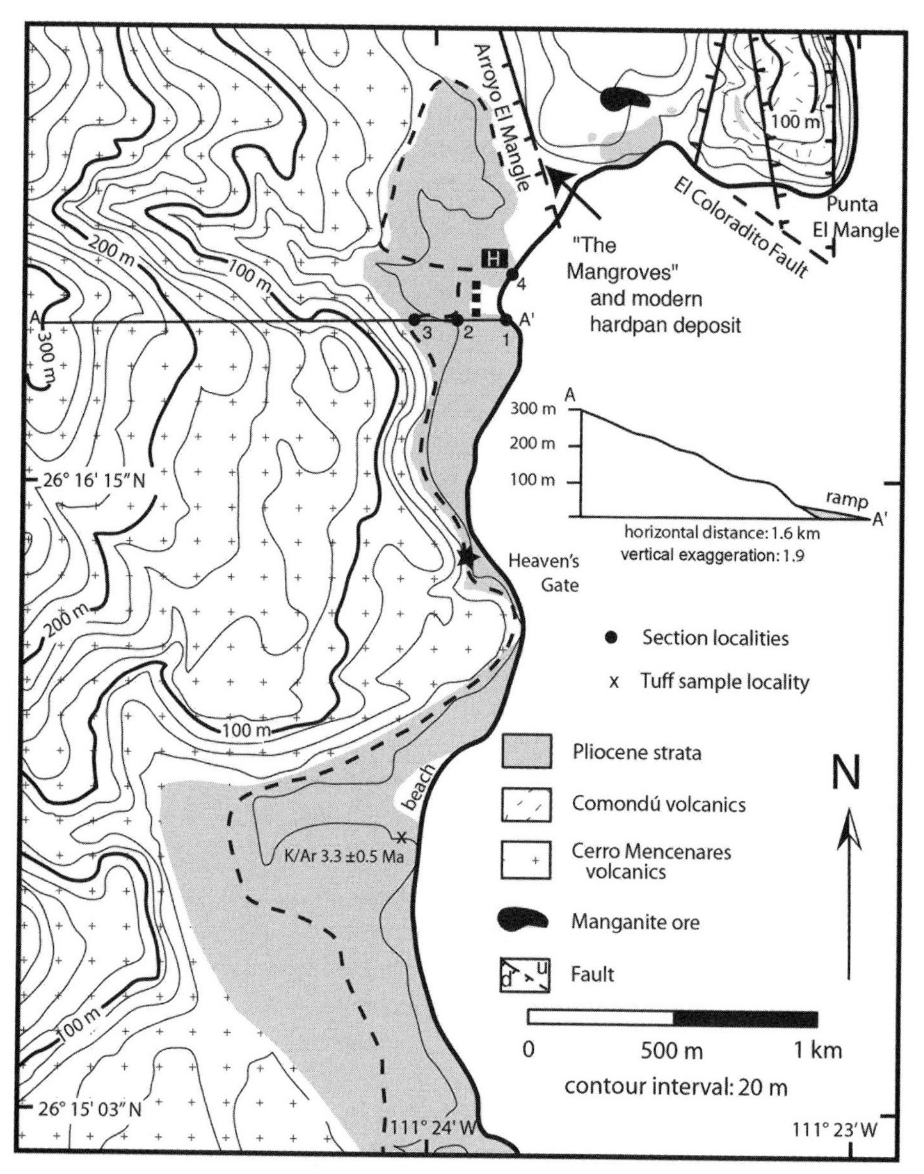

Map 7. Topography and geology of the area around El Mangle. *Note:* Secondary roads and trails are marked by heavy dashed lines; black star marks an entrance gate (Heaven's Gate); black square with letter H marks location of abandoned hotel structure (smaller squares indicate houses); cross-section A to A' marks location of key stratigraphic profiles (1–3) shown in figure 22; location 4 (at hotel) marks another stratigraphic section for comparison (see text). Map by author.

had gone to a great deal of trouble to scrape off the vegetation and thin soil using heavy equipment. The limestone surface was fully laid bare. Shortly after, we reached an abrupt turn in the road, where it descended through a gap in the limestone toward the shore below.

The skeletons of three luxury homes sat at the bottom of the hill near a cobble beach. One was whitewashed and nearly complete with its orange-tile roof securely in place, but another was still only half built, roughly shaped with bare concrete blocks. The third showed only the bare outlines of its foundation. Across the draw on the neighboring limestone promontory loomed the hulk of a steel-frame building. Its pretensions as a hotel were proclaimed by an enormous *palapa* shading a large patio overlooking the sea. Two dogs greeted us, barking excitedly as we drove up to the first dwelling. Soon after, there appeared a thin old fellow, a caretaker, who called off the dogs. His name was Hector. His job was to make sure that construction materials, such as the stockpile of concrete blocks, were not pilfered. On learning that we were harmless geologists, he granted permission to explore the surrounding limestone cliffs. Hector was not very forthcoming, however, in satisfying our curiosity about what sort of outfit was invested in developing the land around El Mangle. We never did come up with answers, even after asking discreetly around town back in Loreto.

Hector gladly accepted some oranges we offered him, and after lunch we set to work checking the limestone cliffs for the fossils that might tell us the depositional environment and age of the formation. Fossil pectens were most abundant, some encrusted by barnacles. We failed to find anything that would delimit the age of the rocks on that occasion. After exploring the area around the largest structure on the property, we located the perfect camping spot for a future expedition. It was on a sand beach to the north, not far from the mangrove enclave that gives the place its name: El Mangle. That was enough for a start; we would be back the following January, in 1999.

During the initial field season, we drove the same track back and forth from San Bruno without too much difficulty. One risky stretch, however, crossed a dust wallow that engulfed our van to the floorboards. Dust invaded every surface, inside and out. The following year, we were determined to find an alternative route to El Mangle. Our crew flew to Loreto, where we rented a light pickup truck. We would have to take two trips to ferry the students and our camping gear from town out to El Mangle. Dave and I left the students in Loreto for a free afternoon and went out to scout a rough road skirting the south margin of Cerro Mencenares.

We passed the new army checkpoint on Mexican Federal Highway 1 at 2 p.m. and informed the soldiers that we were headed only a short piece up the road, where we expected to find our turnoff. It was a challenge, but Dave piloted the pickup over a rudimentary road. Within sight of the coast and our connecting road, we arrived at the edge of a sand-filled arroyo with tire tracks leading across. The crossing looked safe. Dave gunned the engine and we flew into the middle of the arroyo, where the truck promptly sank to its axles. Try as we might to free the truck, all efforts failed. We finally admitted defeat just as the sun dropped below the horizon. Our only option was to reach the checkpoint on foot, five miles (8 km) away, and hitch a ride from there back to Loreto. It was 9 p.m. by the time we approached the checkpoint in total darkness. "*Señor*, where is your truck?" asked a much surprised guard, who remembered us from earlier in the day. Mustering my best Spanish, I tried to tell what happened, shrugged my shoulders, and declared our intentions to head for Loreto.

An officer joined the conversation and the explanation was repeated, once again with more details regarding the spot where we abandoned the pickup. "*No problema*," the officer told us. "We'll get you out." I was certain we would need to hire a farmer with a tractor to recover the truck during daylight hours. What equipment did the army have available? I wondered, so I replied, "*¿Con qué?*" The officer turned and shone his flashlight on half a dozen Humvees lined up in a row. "*Uno de éstos*," he replied with a grin. Thus, we set out to retrace our path under armed escort in a four-wheel-drive vehicle with a motorized winch mounted on the front bumper. I sat up front with the driver to give directions. "*¿Estás seguro?*" he would ask in amazement when I pointed ahead to show where we needed to go. The reaction was one of disbelief that we had manhandled the pickup over the rough road and gotten as far as we did. At last, our lonely pickup appeared in the Humvee's headlights, exactly as we had left it. In no time at all, the winch was hooked up and our vehicle safely pulled out of the sand. To make sure we had no further mishaps, the officer followed us in the Humvee back to the checkpoint. I am forever indebted to the Mexican army for our rescue. We got back to Loreto at 11 p.m., just when our worried students were about to notify the police regarding our disappearance. After consultation with locals the following day, we found a third route that entailed crossing farm fields to reach the connecting coastal road.

Prior to our final field season in January 2002, torrential rains had so destroyed the tertiary roads that we could not drive overland to El Mangle. Indeed, the farm fields we had crossed the year before were completely ruined by a cover of sand and gravel flushed through the distributaries of Arroyo San Juan. We had no choice but to reach El Mangle by boat. On that occasion we were introduced to Leon Fichman (Baja Outpost) and availed ourselves of his services. It was the beginning of a long and advantageous relationship that later led to the exploration of islands in the Loreto Marine Park.

El Mangle turned out to be more complex than first imagined. The landscapes around El Mangle are like a pair of fraternal twins from the same womb: one born a little bit before the other and each with distinct personalities that require patience to know and appreciate. Hence, the following excursions reflect a pair of stories from the area immediately around the unfinished buildings and the area around Punta el Mangle (see map 7). The tales are tantalizing, because they reflect a wider significance related to the separation of Baja California from mainland Mexico and eventual transfer from the tectonic North American plate to the Pacific plate. As such, the tours are modeled on hikes taken in January 1999, when we began to realize the dual nature of the place. Although distances are not great, trying to do the entire configuration in a single day would be overreaching: individual bits and pieces of the puzzle are complicated enough and the terrain sufficiently rough that an overnight at El Mangle is advisable.

FEATURE EVENT

Playing Hopscotch on Tectonic Fractures

Beach camping is an extravagance for those, like myself, deprived of coastal living during a landlocked childhood. After the sun pokes above the open sea on the eastern horizon, it requires mere minutes on a clear morning before strong sunshine illuminates everything in its path. In between is a brief spell under magical light. Dave is up, but he is not gazing out to sea. He stares intently toward Punta el Mangle. I quickly dress for the day. Already, long shadows drape the sea cliffs that project a third of a mile (500 m) eastward to the point. "Strange," Dave confides in me, "just as the sun rose, the cliffs flashed a shade of red." He adds, "There's something over there other than limestone." I have not retrieved my glasses yet, I am too lazy to hunt for my binoculars, and preparations for breakfast must be made. I propose that we keep to our plan. We will return to the limestone sea cliffs near the houses in the next draw south, where unfinished business waits. It is crucial to get a better handle on the biostratigraphic relationships preserved in the limestone beds. Afterward, we can take advantage of a road bulldozed inland around the hotel's construction site to further explore the landscape. It looks like an easy day, perhaps entailing a hike of only a mile and a half (2.5 km). "We'll get to Punta el Mangle soon enough," I promise.

The tide is out. Mustered with full gear for the day, the crew moves south along the cliffs below the main construction site. Still wet from the retreating tide, the limestone bench we cross reveals some whole pectens, many disarticulated shells, and internal molds of various gastropods. Exposed at head level is a thick layer of conglomerate consisting of well-rounded pebbles and cobbles of eroded andesite that mostly come into contact with one another. Pieces of broken pecten shells fill some of the interstices among the igneous clasts. It is a sure sign that the deposit was rapidly dumped into seawater. Limestone layers follow above the conglomerate. High above us, the cliff top appears to be formed by conglomerate (more than 13 ft. or 4 m thick) with andesite clasts the size of boulders. A mental note is made to return another day, when an inventory can be made of all the contents in the various layers.

Beyond the cliff face, a cobble beach extends below the house sites. It is a hard traverse of 500 feet (152 m), before the base of the limestone cliffs on the opposite side of the draw can be reached. It occurs to me that the shore must be fed by a supply of andesite cobbles channeled through arroyos from the volcanic heights above. The rounded beach cobbles look much like those preserved in the cliffs, which we have now passed. During a downpour, the low-lying houses in the draw would surely flood, while the large structure on the cliff top would remain safe. Hector waves at us from the last house. The cliff section now before us is comparable in thickness (about 40 ft. or 12 m) to the mixed conglomerate and limestone layers beneath the hotel. There is, however, a huge difference. These cliffs are dominated by limestone. Some layers are crowded with fossils that verge on the nature of a coquina. It is here that we must hunt in earnest for a fossil to provide a clue to the formation's relative age.

Great blocks of limestone toppled from layers above make an untidy heap at the water's edge, where surfaces remain slippery from the receding tide. Our movements crossing this obstacle are guarded. It is useful to search for fossils in the detached blocks, but to be of any value the fossils must be traced to their original layer in the sequence. The most abundant are separated pecten shells, many of which share the same orientation. Through steady maneuvering, we attain the lowest exposed part of the cliff face where the talus is negligible. Seaward, we reach the front of a massive base (more than 8 ft. or 2.5 m thick) on which the rest of the limestone succession follows (figure 22, section 1). Up to 75 percent of the saucer-shaped valves are deposited concave-side down. It is the most stable final resting position of shells moved postmortem by waves or shifting bottom currents.

Looking like a plump sand dollar, a fine specimen of *Clypeaster marquerensis* is spotted, half exposed in the limestone. This extinct echinoid, with its petal-shaped pores tracing five radiating loops on the aboral surface, is a regional index fossil that denotes a Late Pliocene age according to the correlation scheme devised by the paleontologist Durham (1950), Once the first good example is found, bits and pieces of the same fossil are readily recognized scattered through the outcrop. We busy ourselves with a search among the fallen blocks for a perfect specimen that might be freed whole. The index fossil tells us that the limestone beds at this locality are somewhat older than those above the chert beds on the Concepción Peninsula (see chapter 5) or the sandy limestone deposited within the granite embayment at San Francisquito (see chapter 4). Species that thrive for a time and then are superseded by others in the same genus provide the basic temporal framework against which strata may be compared and correlated throughout the gulf region.

With an important part of the morning's mission accomplished, we rejoin the cobble beach, where attention turns to the cliffs alongside

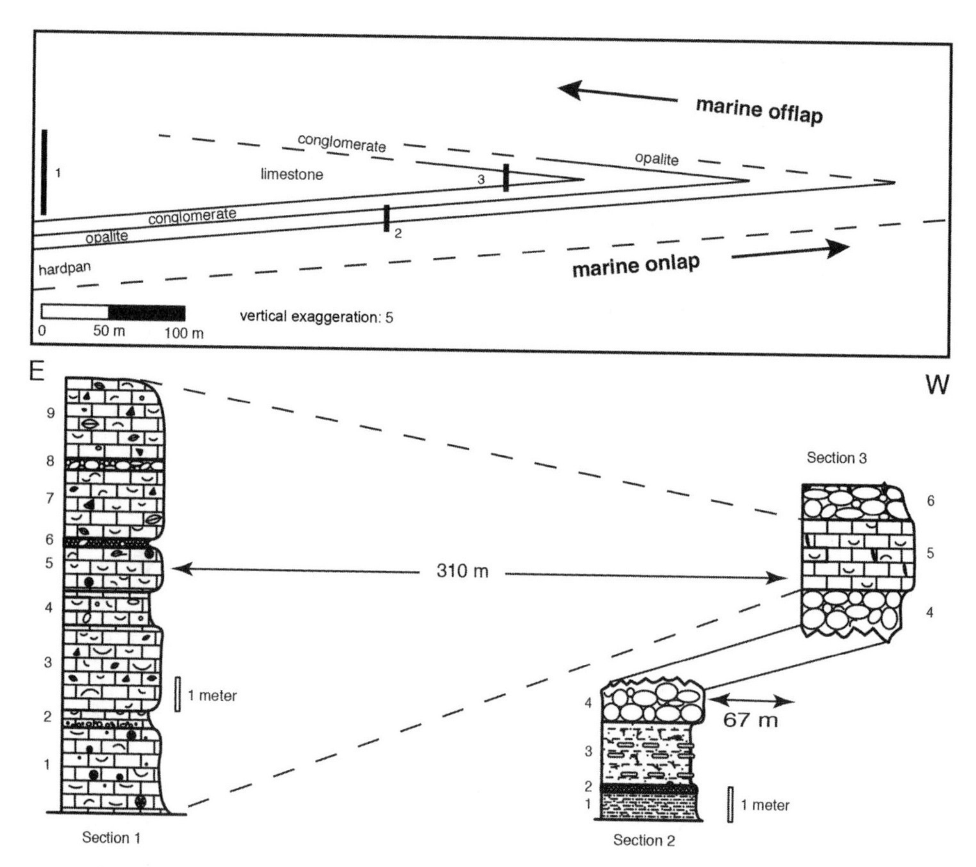

Figure 22. Three stratigraphic columns in a Pliocene sequence (*lower*) correlated in the context of an "Israelsky wedge" that tracks marine onlap and offlap (*upper*). Lithological units in sections 1–3 are distinguished by standard symbols for limestone (brick pattern), conglomerate, and siltstone. Original drawing by author.

the road blasted to improve access to the construction sites for the three dwellings. Connected with the sea cliffs just explored, the road cut adds to an east-west axis stretching more than 1,000 feet (310 m) perpendicular to the margin of Cerro Mencenares (see map 7, localities 1–3 on profile A–A'). The road leading to the top of the limestone embankment cuts through several layers, some of which clearly differ from limestone. At first, juxtaposition of the various rock types is confusing. It takes awhile, walking back and forth, up and down the roughly 330 feet (100 m) of inclined roadway to sort things out.

There is conglomerate, for example, but it is quite different from the fossil-poor conglomerate encountered at the start of the day below the hotel site. There is also siltstone, brownish red in color, with a sizable fraction of clay intermixed. It is not much of a layer (only 2.6 ft. or 0.8 m thick) but is very distinctive.

The key to the riddle is to trace out one of the conglomerate beds. Near the southernmost house, the top layer exposed in the road cut is conglomerate formed by andesite clasts that range in size from pebbles to small boulders (figure 22, section 2). Most of the boulders are covered by a rind of coralline red algae, but otherwise they occur in contact with one another. In places, clusters of barnacles also encrust the algal rind. Among the larger andesite clasts are pockets of limestone rich in fossils that include gastropods and abundant spines from sea urchins (echinoids). The layer represents an open coastal deposit, where boulders were eroded from a rocky shore and gradually reduced to smaller cobbles and pebbles. Frequent turning by waves crashing shoreward enabled coralline algae to coat the smooth clasts on all sides. This habitat was favorable for intertidal to shallow-water invertebrates. The telltale conglomerate forms a layer (6.5 ft. or 2 m thick) that spread landward because of changes in relative sea level that gradually engulfed the former shore.

Not level, but rising with the road, the conglomerate layer is traced inland over a distance of 220 feet (67 m), at which point a second conglomerate layer appears at the top of the road cut, separated from the first by an interval of limestone (figure 22, section 3). The additional conglomerate layer is related to the first, because it signifies a drop in relative sea level and a functional seaward retreat of the habitat zone. In effect, a single conglomerate layer has executed a U-turn, but one that blankets its former advance. Fossil oyster shells are attached to the largest boulders in the upper conglomerate bed, and this is the main difference between the older and younger manifestations of the unit. Continuous limestone layers between the two conglomerate beds record the expansion of a great wedge, apparent when traced seaward to the front of the sea cliffs, where the fossil pectens and associated *Clypeaster marquerensis* are most abundant. Near the opposing apex of the wedge, the intervening limestone features abundant fossil razor clams (*Tagelus californianus*) and small gastropods (*Strombus granulatus*).

Interpreted with some degree of vertical exaggeration, the rock layers connecting sections 1–3 depict a sideways V-shaped arrangement (figure 22, upper part). Stratigraphers know this configuration as the "Israelsky wedge," named in honor of Merle C. Israelsky, a petroleum geologist working in Louisiana during the late 1940s and early 1950s.[3] The concept of different sedimentary rocks and associated fossils as facies contemporaneous with one another from adjacent environments became well established in European academic circles, but the notion was slow to gain widespread use in North America until its utility was demonstrated in the petroleum industry. Israelsky was influential in this regard, but his correlations were drawn from deeply buried strata typically traced over long distances. In a single compact space at El Mangle, the conglomerate and limestone layers with separate fossils illuminate different Pliocene facies exposed at the surface in a well-defined coastal wedge against Cerro Mencenares. Unidirectional limestone ramps (see chapter 4) are commonplace, but those with a pattern of both marine onlap and offlap are unusual. No textbook example based on surface exposures exceeds the clarity of El Mangle in this respect.

Stratigraphic relationships at El Mangle are more nuanced, however, than is shown by the wraparound layer of intertidal conglomerate. Once again, we retrace our steps down the road to fit the distinctive red siltstone layer into context with the rest of the sequence. Alas, without digging we are unable to see what underlies the siltstone. Clearly, when exploring the thickest surviving part of the limestone wedge in sea cliffs to the east (figure 22, section 1), we did so at sea level below our present elevation on the road. This means that the limestone was deposited on an inclined plane as sea level rose during the later Pliocene. In one fashion or another, the siltstone's red color implies that it was derived from a landward source. We must assume that andesite bedrock occurs at some depth below.

Above siltstone but below the intertidal conglomerate with its algal rinds and encrusting barnacles is the most intriguing stratigraphic unit of all at El Mangle (figure 22, section 2). The layer weathers white from rocks that are olive white in complexion in a freshly excavated surface. Discerning continuous sublayers within the unit, which measures almost six feet (1.8 m) in thickness, is difficult. Lenses formed by solid opal trace irregular horizons. Between the lenses is a sandy, silica-rich material. It is poorly cemented and easily disintegrates when struck with a hammer. A hand lens readily reveals that the material includes a significant portion of clay. One after another, members of the crew begin to find fragments of opalized wood. It is an unexpected turn of events. Applying the pick-end of my geological hammer to the outcrop, I dig out a splintered piece of twisted material from a tree limb 3.5 inches (9 cm) in length. The sample retains a surface texture of wood grain swirled around individual knots. Traces of growth rings appear in cross section around the rim, although the core is opaque with shiny opal. The next fragment I retrieve from the outcrop has a core of black opal that preserves distinct growth rings throughout in cross section.

As a paleontologist and geologist, I find two questions beginning to bounce around in my mind. The first, concerning how the Pliocene wood became opalized, is a geological issue, perhaps the easier to consider. Opal is a form of silica, which is readily supplied in solution by geothermal waters. The Gulf of California is a place shot through with fractures, where hot water is plumbed from considerable depths to the surface. A further search along the outcrop brings to light whole chunks of opalized material that replaced dense, mat-like concentrations of woody twigs and stems (figure 23). It seems plausible that woody debris was concentrated in a setting on land but near the edge of the sea. Some time after the deposit accumulated, geothermal fluids saturated and mineralized the wood. We can think of no other term to describe what has been discovered: it is an opalite. It is self-evident that the opalite was hardened by the time the overlying layer of conglomerate arrived. The boundary between opalite below and conglomerate with large clasts of rounded andesite above is sharp and distinct. If not already well consolidated, the layer of woody debris would have been utterly destroyed by wave energy emplacing the conglomerate. Prior to the rise in sea level that inundated the shelf, hot silica-rich water seeped into the deposit of woody debris, perhaps fed by gravity from a nearby spring on the slopes of Cerro Mencenares.

Paleontological in nature, the other question poses technical difficulties: What kind of plants with woody structures ranging in size from twigs to small branches contributed to this unusual Pliocene deposit? In addition, I wonder whether the deposit could include the remains of

Figure 23. Mass of twig-size plant debris from a Pliocene opalite at El Mangle (portion of pocketknife shown for scale). Photo by author.

three-million-year-old cacti. My professional career has been devoted to the fossils of marine invertebrates. While I might qualify for amateur status as a botanist in the most rudimentary sense, my enthusiasm for plants requires extensive help from guidebooks. Surveying candidates from among the existing flora and comparing their woody structures with the opalized fossils might root out the answer. The mere prospect makes me giddy with anticipation. With sufficient fossil samples for the task packed away, we return once again up the road, paying close attention to the layer of opalite. The layer disappears below the road surface before the road gains the top of the hill. Sleuthing around the sharp turn in the road uncovers more evidence of opalite exposed as an artifact of the Israelsky wedge (see figure 22, upper part).

My responsibility is to articulate key questions, such that everyone engaged with the project has the wits to recognize potential answers when these are encountered, but the noon hour has arrived. Students drift down the slope to the junction with a side road that crosses north to the hotel construction site above the houses (see map 7). I am reluctant to attend, because the road south now appears so attractive. The crew is drawn to the hotel palapa, where ample shade is ready. I follow behind, and much to my delight, the lunch spot offers a fine perspective over the road cut with its exposed wedge of strata thrust provocatively against the underbelly of the Cerro Mencenares complex (figure 24). Nature had already carved an erosional gap through the limestone ramp, but somebody excavated a road in precisely the right place to reveal the most salient internal features of its architecture.

Reenergized from the break, we look for a path leading inland from the construction site and upward onto the lower flanks of Cerro Mencenares. Because the site is situated on a limestone foundation, the trail logically must cross the boundary between limestone and underlying andesite sooner or later. Extending a quarter mile (400 m) uphill to the west on a gradient that gradually steepens, the track beneath our feet is gouged from limestone to leave a white streak in the landscape. Passing 65 feet (20-m contour) above sea level, the ground exposed under the sparse desert vegetation on our immediate right and left remains beige and chalky in texture. In the near distance to the south, however, the ground is red. Thus, the landscape clearly mirrors an indentation in the former Pliocene

Figure 24. View looking south over Pliocene strata exposed in a road cut on the margin of the Cerro Mencenares complex (see interpretation in figure 22). Photo by author.

shoreline (see map 7). After a bend in the trail at about 130 feet above sea level (40-m contour), a bulge of red dirt and rocks signals our arrival at the paleoshore. Almost as quickly as it is encountered, we leave the bulge behind and reemerge to the north on chalky ground where the path becomes appreciably wider.

Our unknown engineer failed to scrape an artificial outcrop as deeply insightful as that along the road studied this morning, but the present path strikes perpendicular to the ramp revealed in detail at section localities 1–3 (see figure 24). Enough of the chalky ground is penetrated to plow up fossils from the leading limestone point of the wedge. I stoop to pick up the internal mold of an immature conch (*Strombus galeatus*). It is the same species encountered in the Pliocene outer lagoon on the Concepción Peninsula (see chapter 5), but barely a third the size. A few steps beyond, I spy a broken fragment from the top of a shell belonging to the same species. The piece is large enough to suggest that fully mature individuals of the species (comparable in size to that shown in figure 17) once lived along this paleoshore. It was not smashed apart from a whole shell by the indelicate steel blade of a bulldozer. Bits of calcareous algae coat the coils on the inside part of the shell, showing that the fragment was already broken when cast up by waves on a Pliocene shore.

The path ahead weaves back and forth across a firm boundary etched in rocks between former land and sea. It pushes forward, taking us north by northeast another 1,000 feet (more than 300 m). Abruptly, the entrenched Arroyo el Mangle truncates the landscape (see map 7), and the path takes a sharp turn to negotiate a way to the streambed below. To this point, we have enjoyed beachcombing on a stretch of deserted Pliocene shore well elevated above the present-day gulf waters. It is not some rarified academic exercise but a real and personal transcendence to an earlier time when this majestic gulf was about to undergo major change.

As we descend to the floor of the arroyo, the dark blue waters of the gulf slip from sight. The path ahead crosses the arroyo and loops inland behind the next hill to the north. We leave the path and move downstream toward the coast. The fresh sea breeze we enjoyed on the open slope of the limestone ramp is gone. Air hangs still in the arroyo and reflects hot off dark boulders and cobbles spread among channels closely braided in a streambed divided by clumps of resilient vegetation. Extending behind us, now, the valley that tightly constrains Arroyo el Mangle runs north by northwest along the eastern flank of the Mencenares volcanic complex (see map 6). It is a fault-controlled valley with an orientation similar to fractures previously plotted on the Concepción Peninsula (see chapter 5, map 5). Scattered here and there are pure white pebbles that stand out from the rest like markers in some bizarre board game grafted onto the terrain. The pebbles are made from quartz, eroded from thick veins in the hanging walls of the valley and transported downstream by flash floods. There is a tectonic overprint to this place that shouts out a story of inelastic rocks strained beyond the breaking point.

Near the shore, the valley opens on a web of greenery dominated by salt-tolerant trees such as the white mangrove (*Laguncularia racemosa*) and shrub-size saltbush (genus *Atriplex*). We have reached the heart of El Mangle, where mazelike thickets of head-high vegetation are interspersed with dark pools of stagnant water. Knowing that we will want to end the day at camp on the beach below the hotel construction site, we make every attempt to steer a southerly course through and around the tangle of vegetation. At the side of El Mangle, an open floodplain lies crusted with sediment that appears gray on the surface. Moved by impulse, I plunge the pick-end of my geology hammer through the crust into the dirt below, giving it a stir. The soil is brown-red in color from the silt and clay eroded from andesite hills. The coastal plain deposit on which we squat looks every bit like the Pliocene siltstone layer in the Israelsky wedge only a third of a mile (500 m) to the south. Moreover, the floodplain is strewn with plant debris, dry and twisted branches woven together along with piles of twigs left by the last flood. Here, then, we have an apt Lyellian model to apply to the stratigraphic puzzle.

After swinging my backpack to the ground, I retrieve the largest piece of opalized wood from a secure compartment. With the exhumed piece in hand, I stride back and forth from one pile of naked brush to another. I am searching for a stick of comparable size that will exhibit similar wood grain and knotty texture. There is much to choose from, and I resort to breaking more than a few dry branches to stuff the samples into my shirt pockets. The crew fans out to help in this endeavor, and I ask that attention be paid to even the smallest stems and twigs. The exercise is easy enough, but now comes the hard part, when the worn bits of fossil woody debris must be matched with live samples for identification. No living cacti are within close range, a fact I observe with regret. Some of the crew peel away and return to camp to organize the evening meal. When I turn in for the night, my brain is flush with images of three-million-year-old fossil wood laid side by side with bits of woody brush that were green with life no more than a year or two ago.

On day two, the crew assembles for a frontal attack on the headland of Punta el Mangle. It will not be easy. The first obstacle on the march north is to fight beyond El Mangle (the mangroves). Taking the path of least resistance, we detour to cross the arroyo inland at a spot relatively free of plant life. The maneuver brings us squarely against a ridge 130 feet (40 m) high, which we skirt around south and east toward the coast. Unhappily, the wash of the open streambed does not favor this side of El Mangle. Plant life crowds the north wall of the arroyo, making it a struggle to reach the shore. There, we find ourselves back in the domain of limestone sea cliffs (see map 7). The bluffs along this part of the shore, however, are different. It is not so much that the tidal zone is chaotic under a jumble of fallen blocks. The going is slow, but all obstacles are passed with due care. What is odd is that many of the fallen limestone blocks are transformed to brittle, porcelain-like masses with sharp edges.

In places, the limestone is bizarrely discolored rust orange. Fossils are present but barely recognizable. It requires some imagination to recognize them, but large fossil oysters occur intact (articulated) as part of what was an oyster reef. Elsewhere are distinct fragments of broken corals. Much of the limestone along this stretch of the coast appears to have suffered alteration from the seepage of hydrothermal water. After hard progress, we emerge from below the limestone cliffs, exiting to a cove at the corner of the bay with a small, northward-pointing dale.

Rocks exposed east of the cove are not limestone, but close investigation is barred by sea cliffs that plunge vertically into the water. No available space exists to crawl along the side of these sheer cliffs. Low ground at the back of the inlet offers an alternative path through the steep dale, giving access to the heights west of Punta el Mangle. As we climb on a tangent to the slope, the first 130 feet (40 m) are a punishment. Thereafter, the slope flattens on the way to an elevation more than 330 feet (100 m) above sea level. A ledge of limestone only 32 inches (1 m) thick is exposed along the curve of the hill in our path (see map 7). It contains large fossil oysters and distinct coral fragments reminiscent of those from the limestone cliffs below. Here, the limestone sits uncomformably on volcanic breccia. This relationship suggests that the valley wall we have just negotiated is on the upthrown side of a fault. Presumably, limestone of a thickness equivalent to that of the cliffs below once blanketed the fault block on which we stand. Another fault shows itself farther along below the top of the hill. The relationships suggest that the surviving limestone forms the cap of a narrow horst.

From the crest of the ridge, the view southwest is oblique and somewhat abrupt, as might be seen from a small aircraft. The scene takes in El Mangle and all beyond to encompass the enormity of the carbonate ramp with its stunning Israelsky wedge (see plate 8). The sea breeze is refreshing, and turning one's back on the picture-book view is difficult. The task demanding immediate attention, however, is to explore the ridge and find the blush-red rocks that appear on the seaward face observed by Dave on our first morning.

We spread out to canvass the rest of the ridge above Punta el Mangle. A third fault, crudely parallel to the others, cuts along the outside lip of the ridge overlooking the sea to the east (see map 7). The relationship is confirmed through repeated perambulations, back and forth. The space between the outer faults is confined to the ridgetop, and the interval so defined represents a small graben. It is odd that no limestone is captured in the graben. A systematic search in the immediate area finds no trace, so we return to the nearby limestone outcrop on the other side of the hill for more clues. It is slablike in dimensions and tilts eastward like the master ramp with the Israelsky wedge to the south, but more steeply inclined. The clinometer on my compass shows that the miniramp has a dip of 12°, far too much for an undisturbed structure. Apparently, the miniramp was oversteepened sometime after it formed. The relative steepness of the dip angle also implies that the limestone layer vanished to the east.

Attention shifts to the color of the sea cliffs below. Moving down the slope beneath the limestone to the edge of the cliffs, we reach a sheer face of polished rock that shines almost scarlet in the midday sun. A hint of grooved striations is reflected from the surface.

The markings are indicative of friction between two rock bodies that moved in opposite directions on opposite sides of a fault plane. *Slickenside* is the term applied to fine, parallel scratches left in the opposing rocks as a result of the violent snap. Bemused, we find ourselves standing next to another fault, but it is a major one with a radically different orientation to the others. The surface color is due to hydrothermal baking of the fault zone, because it fades through maroon to dull olive gray beneath.

The fault plane on the sea cliff is precipitous. Closer examination of the surface is too dangerous. Much of the morning, we have played a game of hopscotch, crossing a web of intersecting faults. Here, the stakes in the game rise markedly. This particular fracture is significant, because it separates the undisturbed master ramp stretched before us to the south (see plate 8) from everything else to the north. Dave gingerly crawls forward to the edge and sights with his compass to take relevant readings. The fault plane shows a strike of N 55° W, an orientation that extends landward to join at an obtuse angle with the normal valley fault that we crossed earlier in the day (see map 7). A natural extension in a straight line eastward passes out to sea. Our scarlet fault plane dips steeply 54° to the south. We are seated near the southwest corner of a substantial fault block raised above the level of thick limestone layers exposed elsewhere in the surrounding landscape. It is urgent to find a way down the cliff to a place where we might see strata hidden by the fault plane.

The lunch break provides an excuse to back away and calmly reassess the situation. To reach the rock layers behind the distinctive El Coloradito fault plane, we track along the south rim of the bluffs, keeping as low as possible, 130 feet above sea level (40-m contour), where it is safe to walk. Members of the crew are alarmed at the initial prospect, but a kind of channel in the rocks provides a passage all the way to the water's edge. Dave and I descend first, taking care to shuffle downward in a seated position facing outward with knees bent forward and boots flat on the rock surface. The rest of the crew follow after in pairs. I return topside to guide the last, hesitant member down. A rock face is accessible along the shore to the west. It is exposed beneath an overhang that retains the polished fault plane. Here we have a decent place to start the accounting. Other layers higher in the section are well exposed in the channel we have descended.

The west end of the tectonic block exhibits its own unique stratigraphy, unlike anything else in the area. Volcanic breccia forms the base, containing soft aggregates of olive-shaded minerals set in a hard maroon matrix. Scattered throughout are some fossils: small clams, gastropods, and rare pectens. Clearly, the breccia was ejected into seawater. Overlying are interbedded layers of sandstone and mudstone. These show well-defined bedding planes that allow for strike and dip to be measured. Most instructive, the layers conform to a strike different by 35–40° from the general north-south orientation of the grand limestone ramp to our south. Their inclination, 25–35° E, also is much exaggerated in comparison with the limestone ramp. Above is columnar basalt overlain by more volcanic breccia. The package comes to a total thickness of about 277 feet (115 m). The only visible overlap with the neighboring ramp to the south is the thin cap of limestone preserved on the west side of the ridge above. As andesite is the prevalent igneous rock exposed to the south, we must entertain the interpretation that other layers of basalt and volcanic breccia (comparable to Comondú volcanics) occur at a depth beneath the andesite. In all its glory, this El Coloradito Fault provides considerable insight on the nature of rocks concealed, elsewhere, below the surface. At minimum, about 230 feet (70 m) of uplift occurred along the fault plane.

Punta el Mangle is close by. Little talus has collected at the bottom of the cliffs to block the way, and the going is easy. Rocks exposed on the point provide a totally different impression of the geology on the north side of El Coloradito Fault. This is because our path crosses from horst to graben and back to another horst. All are defined by older faults truncated by El Coloradito Fault. Exposed at the point is a gigantic pile of volcaniclastic rocks, crudely stratified with thick layers of andesite boulders separated by lesser seams of volcanic ash. The most telling lesson of the day is that the greater fault block consists of discrete segments first brought side by side by the jostle of north-south-oriented faults prior to being decapitated by the younger master fault: El Coloradito. The key question to ponder is what induced such a contrary fault trend in El Coloradito compared to the others, including those that demarcate the narrow El Mangle valley.

We have reached the farthest point on the day's tour. While the distance on a diagonal route across the water is not so much, we must return more or less the same way we arrived. It is a matter of reverse hopscotch: jumping across side faults on the way back to the chute, where we climb upward perpendicular to the fault plane of El Coloradito. Safely onto the ridge, one more fault remains to cross in order to descend to the inlet at the inner corner of El Mangle Bay. Here, we may backtrack along the same laborious path at the water's edge, or we can circumvent the coast and the mangrove swamp by climbing inland to explore the upper limestone bluffs. As there seems to be no stomach for talus jumping at the shore, we take the climb westward through limestone layers. A bit less than 100 feet (or about 30 m) above sea level, we encounter manganite ore eroded from the limestone. It is a sizable ore body. Pieces as large as small boulders sit loose at the surface. The ore is distinguished by several traits that include color (iron black in this case), hardness (4 on Mohs scale of 1 to 10), and specific gravity (4.3).[4] Picking up a sample (more impressive than any from San Francisquito) is hard to resist, but the ore is heavy.

Turning to face the water, becalmed under the shelter of El Mangle Bay, we cannot overlook the fact that the ore body is aligned on a direct path with El Coloradito Fault (see map 7). Mineralization of this sort is characteristic of fault-hosted deposits associated with low-temperature hydrothermal solutions. Following the precise fault trace is difficult, but El Coloradito appears to extend inland and intersects the hanging wall of the Arroyo el Mangle Fault. We have stumbled onto a major nexus of secondary mineralization related to the complicated fractures shot through the Gulf of California and adjacent Baja California peninsula. It is not about fossils or stratification, but the thought adds to the intellectual baggage we carry back to camp.

Some fieldwork seasons are more extreme than others, but all come to an inevitable conclusion when it is time to pack up tents and return home. Commonly, it takes me days to readjust to my normal surroundings of classroom, office, and domesticity. By night, my dreams are filled with the bright sunshine and earthen tones of Baja California set against the azure of the nearby sea. When I awake, I am disoriented to find myself in a comfortable bed. By day, team members make time between other responsibilities to study the rock and fossil samples we have collected, to make new maps, and to scour the maps made by others for clues to unravel a landscape's intricate history. El Mangle held our attention through several years, coming to exert an abiding presence heightened through repeated visits. Heaven's Gate not only opened to a core area of unusual beauty and multifarious natural history but also begged our indulgence to understand its relationship with the surrounding region.

Our El Mangle report appeared four and a half years after we first set foot on the place (Johnson et al. 2003). A critically important clue from just outside Heaven's Gate was a tuff layer (volcanic ash) discovered by Jorge below the limestone ramp (see map 7). Under radiometric analysis of potassium and argon isotopes, the tuff yielded an absolute age of 3.3 ± 0.5 Ma. It means that the overlying limestone units wrapped in the Israelsky wedge postdate that time and can be assigned to the lower part of the Piacenzian Stage in the Pliocene.[5] It is an attribution consistent with the relative age set by Durham (1950) on the basis of the index fossil *Clypeaster marquerensis*. Another embellishment of the story came during our final field season, when we were forced to travel by boat from Loreto to reach the study site. The boat trip gave us the opportunity to view rock layers exposed in sea cliffs north of Punta Boca San Bruno. We learned that the brownish-red siltstone at the base of the stratigraphic sequence together with the overlying opalite are exposed in continuous beds traceable over five miles (8 km) to the south boundary of El Mangle (see figure 22). The prospect remains to explore this part of the coast more thoroughly in search of Pliocene wood from other than the saltbush (genus *Atriplex*).[6] Hope remains that fossil cacti may yet be found.

Another boat trip took us from El Mangle north along the coast to Punta Mercenarios below San Basilio (see map 6). That excursion gave us the chance to appraise the complicated structure of the tectonic El Mangle block offset from the Cerro Mencenares volcanic complex. The topographic map for the region lends emphasis to the narrow fault valley separating these two features and the obtuse angle with which it meets our El Coloradito Fault. The latter, with its memorable blush-red continence (see plate 9), cuts the carbonate ramp (see plate 8) and postdates the history of marine onlap and offlap captured therein. The last piece of the puzzle fits into place not through additional fieldwork but through related office work. Transform faults and fractures in the modern Carmen and Farallón basins within the Gulf of California share a similar orientation (around N 55° W) with El Coloradito.[7] The big maps interpreting the submarine geology and bathymetry for the Gulf of California tell a story of ongoing transtensional stress released through a series of segmented transform faults offshore Cabo San Lucas to the delta of the Colorado River and beyond through the San Andreas Fault all the way to San Francisco in Alta California.

According to Umhoefer and colleagues (1994) and Dorsey and colleagues (1995), the South Loreto basin (see map 6) developed under transtensional stress coincidental with the deployment of the boundary between the Pacific plate and the North American plate inside the Gulf of California about 3.5 million years ago. Prior to this pivotal change in tectonic regime, marine basins in the proto–Gulf of California were enlarged through simple extensional rifting of the crust between mainland Mexico and the Baja California peninsula. Normal faults with relative up-down movements were the result. A series of parallel, normal faults above the rift led to differentiation between neighboring horsts and grabens. The wedge of early Late Pliocene limestone at El Mangle was deposited in a period of relative quiet after the last stages of simple rifting in the proto–Gulf of California.

When the East Pacific Rise nosed into the proto–Gulf of California and introduced tectonic forces associated with transform faults, plate subduction on the west side of the Baja California peninsula ceased. Relative movement on transform faults is typically lateral, resulting in offset of points formerly side by side from one another on the same fracture. El Coloradito Fault records uplift on its north side, but its orientation is consistent with a fracture related to a transform fault activated sometime after the nearby tuff bed was deposited at 3.3 ± 0.5 Ma. Some fractures of this kind cross into the Baja California peninsula, but El Coloradito was stopped in its tracks by the massive Cerro Mencenares volcanic complex. It is pure speculation that uplift occurred because of buckling of the crust when Cerro Mencenares rebuffed lateral movements on the fracture. The conundrum is something I am reminded of when the huge, distinctively radial edifice of Cerro Mencenares (see plate 10) looms outside the window as my flight begins its descent to Loreto. When the viewing angle is right, the entire coastal stretch from Punta Mercenarios to Punta el Mangle is visible for a few fleeting moments.

7

Coral Reef on a Volcano at Isla Coronados

> The natural features of small islands provide an ideal setting for imaging small worlds.
> —*John Gillis,* Islands of the Mind

PACKING EVERYTHING WITHIN boundaries made finite by the surrounding sea, islands are famously practical subjects for field studies. Because they are ready-made as discrete packages of geography, any brand of research reaches its absolute limits when nothing else on an island remains for further examination. The smaller the island, the easier it is to manage. There are 40 named islands in the Gulf of California: they come in all shapes and sizes, ranging from Isla Tiburón with an area of 472 square miles (1,223.5 km^2) to tiny El Cholludo with an area of 0.2 square miles (0.05 km^2).[1] Most can be viewed from vantage points along the shores of the Baja California peninsula, although some, such as Isla Angel de la Guarda (see chapter 2), are large enough to be mistaken by the uninitiated for the Mexican mainland on the opposite side of the gulf.

Isla Coronados (see map 1, locality 7) is the most perfect island in the Gulf of California. It is small in size, covering only 2.9 square miles (7.5 km^2), and projects the classic silhouette of a volcano when viewed from any quarter. In fact, Isla Coronados is an extinct volcano that was last active only about 160,000 years ago (Bigioggero et al. 1988). Visitors to Loreto will instantly recognize the profile, as seen from the town harbor 7.5 miles (12 km) away. The adventurer may approach much closer to Isla Coronados without actually setting foot on the island by taking a dirt road to its terminus about 5.6 miles (9 km) north of town at Punta el Bajo de Tierra Firme (see map 6). The closest vantage point from the road is 1.6 miles (2.6 km) due west of the island.

Prior to my first landing on Coronados, I stood at Punta el Bajo looking across the channel to the island on countless occasions. It was largely a question of money that kept me from getting there earlier. Any boatman in the Loreto harbor will gladly take you to the island, once a suitably remunerative deal can be struck. Our annual excursions to Baja California were supported by lean budgets even during the best of times, and cash was dispensed down to the wire by the end of each trip. It did not cost much to drive out to Punta el Bajo, and the view across to Isla Coronados grew ever more enticing (figure 25). Beyond the island's classic profile, it is child's play to recognize color variations that outline the youngest lava flows that spill down the side of the structure.

Attractive as volcanic features are, other aspects of the island's physical geography grabbed my attention. The island's western beaches, for example, are white with sand derived from organisms that secrete calcium carbonate. I suspected that rhodoliths (unattached, coralline red algae) might account for a significant component of the beaches, because we sometimes found the spherical forms washed up on the beaches near Punta el Bajo. Hurricane Marty, which entered the Gulf of California on October 22, 2003, was responsible not only for damaging the Loreto harbor but also for sweeping a mass of rhodoliths into the supratidal zone near El Bajo. Underwater survey work by marine biologists (Foster et al. 1997) has shown that the channel between El Bajo and Isla Coronados is a place where rhodoliths thrive today at depths of 40 feet (12 m) or less. Modern sediments washed onto the beaches of Coronados were making future limestone formations, and I was certain that limestone layers had to exist farther inland as a testimony to bygone times when the sea level stood higher than it does today.

Viewed from the shores of El Bajo (see figure 25), the odd asymmetry of the island suggested that the gentle slope extending south well beyond the base of the volcanic cone might be related to a carbonate ramp. Try as I might with my binoculars to discern any sign

Figure 25. View of Isla Coronados looking east from Punta el Bajo. Photo by author.

of sedimentary rocks layered against the south flank of the volcano, nothing seemed definitive in this regard. Even so, the abrupt sea cliffs and jagged sea stacks cut in volcanic rocks on the north end of the island bore mute witness to the fact that erosion was the dominant theme on the windward coast. It made sense that carbonate sediments had a better chance to be left in place to form limestone in a more protected, leeward setting. Theoretically, I was correct in these assumptions, but the gentle slope in the profile so prominent on the south side turned out to be false in shaping my expectations. Once planted, nonetheless, the seed of an idea grew in my imagination until it became an obsession. There was nothing else to do about it. I had to get myself to Isla Coronados, where on arrival the first thing I intended to do was to locate the southern carbonate ramp.

Some landscapes we have visited are so far off the beaten trail that we fancy ourselves the first to consider them in a systematic and scientific way. The same cannot be said of Isla Coronados. Many naturalists and geologists preceded us to the island. Some left behind detailed observations (Durham 1947, 1950; Anderson 1950). Others, still very much alive, are capable of sharing their own impressions of the place. We went to Coronados, however, with fresh eyes and a determination to find on one small island a set of recondite patterns left in plain sight, yet illustrative of principles regulating natural commerce across the Gulf of California as a whole. Such an obstinate streak of originality is part of the human spirit. It is a trait that allows us to enjoy the thrill of seeing (and understanding) something for the first time, although others have trod the same path before us.

The opportunity to spend two nights on Coronados arrived in January 2004 at the close of an arduous field season on Isla del Carmen. Three of us landed on the island late in the day: Dave, myself, and our student Mike Eros. Scarcely enough sunlight remained for us to set up our tents. We did so on the neck of land that pushes west from the south shore to make a great hook sheltering the largest bays on the island (map 8). A cruise ship catering nature tours was anchored in the bay off the central beach. We felt self-conscious, not wanting to reveal our presence. For one thing, our personal hygiene was not in the best state for polite social engagement, as we had spent a couple weeks on Carmen without recourse to hot bathwater. For another, it was a tenuous proposition that our research permit for Carmen extended to nearby Coronados. All islands in Loreto Bay National Park are regulated under a host of detailed restrictions.

Awake at sunrise, we were relieved to see that the cruise ship was gone. We seemed to have the island to ourselves. After a light breakfast, the three of us set out on separate hikes. Mine took me due east along the south shore on a predetermined quest to find the island's great limestone ramp. In this, I was soon to be corrected. The trek that materialized, however, took me on one of the most exhilarating passages I have ever made as a sojourner through ancient landscapes. When we reunited at lunchtime, Dave and Mike were enlisted to make the journey, as well. Immediately thereafter, we decided to bring a sizable team to the island the following January for a multipronged attack aimed at general mapping and survey work. Required research permits had to be obtained from park authorities for a party of three professors, one graduate student, and a dozen students to spend a week on Isla Coronados. Three reports eventually were published to give the results of the 2005 expedition (Johnson et al. 2007; Ledesma-Vázquez et al. 2007; Sewell et al. 2007).

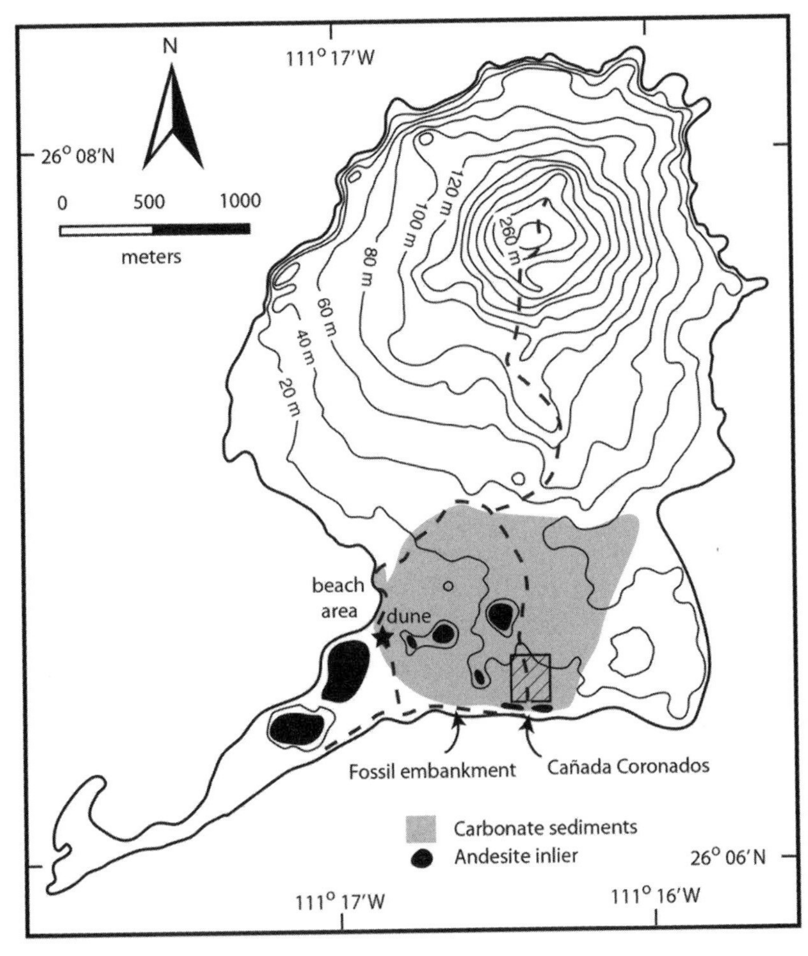

Map 8. Topography and geology of Isla Coronados. *Note:* Heavy dashed line marks island trails; black star shows location of a dune field; box with slanted lines gives the dimensions of Cañada Coronados exposing a fossil coral reef. Map by author.

The project became the template for four subsequent visits to capitalize on the perfect outdoor classroom for students, alumni, and others in January/February of the years 2007 to 2011. Herein, two hikes detailed for Isla Coronados are the most tested of all those featured in this compendium. They may be completed on a single day-trip from Loreto. Visitors must remit a small fee for a day pass to Loreto Bay National Park. Nothing (sand, shells, flowers, fossils, volcanic rocks) may be removed from the island. Shade is provided by two large palapas on the central beach. Sanitary facilities (a pit toilet) are available.

FEATURE EVENT

Transect Across a Fossil Reef and a Climb on a Volcano

Our boatmen stand ready to cast off from the Loreto harbor to Isla Coronados at 7:30 a.m. Other boats were out on the water for fishing earlier in the day and have returned to home base. Brown pelicans (*Pelecanus occidentalis*) roost quietly on huge blocks of rock forming the harbor's artificial breakwater. They are sated, having consumed easy spoils tossed into the water from fish-cleaning stations. It is an ordinary January morning with no clouds overhead and a clear horizon to the north. We have arranged for two boats. Food chests and hiking gear for the day's excursion are quickly loaded. Life jackets secured, we slowly exit the southeast corner of the harbor. Taking a hard left turn, the boats line up on a northerly course for the island. It is all but impossible to carry on a conversation above the whine of the engines, but the boatmen are alert for wildlife. It is too early in the season for the blue whales (*Balaenoptera musculus*), but I have seen humpback whales (*Megaptera novaeangliae*) this time of year. Killer whales (*Orcinus orca*) are a rarity hereabouts, but pods enter these waters from time to time, and I have witnessed their curious approach to boats off the south end of Isla del Carmen. Nothing much stirs this morning. Twenty-five minutes out, we make the wide turn around the slender arm of land that forms the western extension of Coronados. As we glide toward the white beach of the inner bay, the water changes from cold blue to the warmer aquamarine reflected off the shallows.

A current of excited anticipation registers as something more than the genial expectations for a fine day. Arrival on Isla Coronados is akin to entering an island paradise, albeit one without luxurious vegetation in the form of swaying palm trees. The beach could not be whiter, the water could not be more inviting, and the hulk of a Pleistocene volcano hovers above it all like a guardian spirit. Disembarkation is quickly achieved, and our food chests are transferred to the shade of a commodious palapa. The boatmen will return at 5 p.m. for the trip back to Loreto.

Some make it their business as geologists to study beaches as a full-time occupation. I am not one of them, but the idea never ceases to impress me. Why shouldn't work be both serious and fun at the same time? This beach is a rarity well worth the trouble to study, and the crew gathers around for a quick lesson on beach composition. Kneeling above the gentle swash of the waves, everyone takes up a dry handful of sand and examines the particles. It is easy to spot larger fragments of coralline red algae from broken rhodoliths. They are twig-like in shape, with fragile branching bits bleached white from their original rose pigmentation by the strength of the sun. Specialists who study these particular algae describe their growth as fruticose, in reference to the swollen, bud-like tips at the end of delicate branches. Other kinds are referred to as lumpy, having far more robust branches with knobby terminations.

It is true that detritus other than that from rhodoliths litters the beach. Some unpaired shells lay scattered in the sand. They are derived from small, bivalved mollusks. They stand out, because they are familiar to us. Other fragments derive from the broken spines of sea urchins. Some black particles also are a component of the sand. These dark grains were eroded from nearby volcanic rocks along the shore. On the whole, however, the little shells, spines, and bits of andesite constitute only a small fraction of the beach sand. Using a hand lens, one can look more closely at individual sand grains and recognize that many come from the bud-like broken tips of fruticose red algae. Seldom more than 0.06 inches (1.5 mm) in diameter, these small grains are superabundant.

The soft lapping rhythm of the waves is beguiling. Far more energetic waves introduced the rhodoliths to this shore during events when strong winds funneled through the channel between Coronados and Punta el Bajo on the peninsular mainland. Sea swells travelling down the west coast of the island are turned by refraction to enter the embayment and are transformed to powerful surf in the shallows. Rolling naturally over the seabed, rhodoliths are jostled more violently against one another under these conditions. They pile up in the shallows, where friction plays a greater role. Repeated impacts lead to progressive fragmentation. Even a hard cobble of basalt becomes reduced to a pile of pebbles under continuous assault by crashing waves. The porous rhodoliths are readily crumbled against one another during such an onslaught. A beach accrues from the materials that are available. Here, rhodoliths are the most common sort of marine organism living in the adjacent waters. One could say that the rhodoliths are harvested from time to time to renew the beach with a fresh supply of carbonate sand.

Our morning's tour is under way, and the next stage is to advance south toward the other palapa at the opposite end of the beach. We pause there and examine the beach sand again to verify that rhodolith detritus reigns supreme. More white sand rises three feet (1 m) or so behind the palapa. It belongs not to the beach but to a coastal sand dune. We traverse the dune inland to intersect a well-traveled trail that leads to a hillock marking the dune's forward crest (map 8, star). It was here that geology student Ashley Sewell conducted an elaborate experiment to determine the richness of rhodolith material sequestered by the dunes as a result of beach deflation induced by the stiff winter winds (Sewell et al. 2007). The dune covers about 3.7 acres (15,000 m^2), and we stand nearly in the middle at its highest elevation.

Conducting multiple trials, Ashley used a microscope to sort random samples of dune sand according to the origin of the grains. She found that the dune sand consists of rather smaller grains than is found on the beach; it exceeds a value of 96 percent bioclasts derived from mollusks, foraminifera, sponges, sea urchins, bryozoans, and rhodoliths. Among these, the bits from crushed rhodoliths are dominant, at 86 percent of the whole. In contrast, the inorganic contribution to the dune from dark minerals and silicate grains amounts to only 3.5 percent. Further experiments were conducted to relate the dune's total volume (about 39,238 cu. yd. or 30,000 m^2) to the number of pulverized rhodoliths that would be required to fill 86 percent of the dune's volume. Clearly, the raw number of rhodoliths necessary for the job would depend on the average size of rhodoliths in the channel between Isla Coronados and Punta el Bajo. Based on a rhodolith model with a hypothetical diameter of two inches (5 cm), the number of rhodoliths needed to construct the dune comes to nearly one-half billion. There are many coastal sand dunes distributed along the peninsular shores of Baja California and its related gulf islands (Backus and Johnson 2009), but so far, the Coronados dune is the only one recognized as overwhelmingly formed by deconstructed rhodoliths.

We move off the dune crest to the bottom of the leeward slip face to find ourselves in a hollow sheltered from the breeze. The cross-island trail (see map 8) continues from this point and reaches the south coast after a quarter mile (380 m). Upon climbing out of the hollow, we are soon rewarded with a fine vista over the Carmen Passage between the peninsular mainland (west) and Isla del Carmen (east). The channel rolls out before us like a great, blue corridor. From where we stand between inliers of andesite basement rock on this part of the island with the volcano at our backs, we can clearly see that Coronados is a colossal plug at the far end of the long, north-south strait. The depth of water in the passage is considerable, more than 200 fathoms (1,200 ft. or 365 m).[2] It means that much more of Isla Coronados than meets the eye lies below the surface of the sea.

Limestone deposits are exposed beneath low bluffs at the edge of the water. We take a side excursion for a quarter mile (380 m) to the west (see map 8), stopping now and then to examine the outcrops. The most peculiar-looking fossils I have encountered on Coronados (seldom observed in other parts of Baja California) await our attention. They are partly eroded but represent a semi-infaunal species of clam preserved in growth position. Reminiscent of holiday luminaries, paper bags filled with sand that anchor a candle, these eroded fossils (figure 26) are ancestral to the pen shell (*Pinna rugosa*) typically found in sheltered bays up and down the Gulf of California. Elongated with a thin shell, the bivalve lives upright in the sand with only the top part of the fan-shaped end protruding above the surface. They are locally harvested in large numbers as shellfish and commonly appear on restaurant menus as scallops, although the Spanish name (if used at all) is *callo de hacha*.[3] The fossil before us (figure 26), however, belongs to the extinct species *Pinna corteziana* (Durham 1950, 57). It is enormous and must have yielded a large and tasty clam steak, although no one was here in the later Pleistocene to enjoy them. Then, as now, the island's south side was more sheltered from waves than on the north exposure.

We return eastward along the bluff top and soon rejoin the trailhead. Vegetation is sparse, and progress along the bluff is easy. There is no marked trail, but a faint track exists, and the way east is open for

Figure 26. Fossil pen shell (*Pinna corteziana*) preserved in growth position. Photo by author.

the next 350 yards (0.3 km). Bushes and thorny brambles throng an arroyo, which we descend into from the bluff to cross to the opposite embankment (see map 8). Below the embankment, the ground where we stand is littered with a profusion of Pleistocene shells that weathered loose from the cliff face and tumbled down the slope. The assemblage represents a cornucopia of marine invertebrates that once thrived on the south coast of Isla Coronados. Members of the crew pause to separate the shells into small heaps of like species that reflect the astonishing range of the mollusk tribe.

In surveying the fossil deposit, we take the pulse of a former seabed in terms of species diversity and relative abundances. It is an exercise that, because of the munificence of the place, requires little

TABLE 1. Pleistocene bivalves (phylum Mollusca) from the south side of Isla Coronados

Species	Preservation	Dominance
Anadara multicostata	W	A
Barbatia reeveana	D	R
Cardita megastrophia	W	C
Chione californiensis	W	C
Codakia distinguenda	W	C
Crassatellites digueti	D	R
Divaricella lucasana	D	R
Glycymeris maculata	D	R
Lyropecten subnodosus	W & D	A
Lima tetrica	D	R
Megapitaria squalida	W & D	A
Modiolus capax	D	R
Ostrea fischeri	D	R
Pecten vogdesi	D	R
Periglypta multicostata	D	C
Spondylus crassisquama	W	A
Spondylus victoriae	W	R

NOTE: In terms of preservation, W = whole (articulated), D = disarticulated. With respect to rank dominance, A = abundant, C = common, R = rare.

time. In short order, we tally 17 different bivalve species (table 1) and 20 gastropod species (table 2). In addition to the mollusks, there are two kinds of sand dollars (*Encope californica* and *E. grandis*), as well as many coral colonies (*Porites panamensis*). Virtually all shells are preserved in good condition (unbroken) and represent a range of different habits. Among the largest gastropods, the tulip shell (*Fasciolaria princeps*) is a carnivore, while the conch (*Strombus galeatus*, previously encountered in chapter 5) is an herbivore that consumes algae. Nearly half of the bivalves are whole (articulated), but they also represent a range of different lifestyles. Some (*Barbatia reeveana* and *Modiolus capax*) live attached to rocks, while others (*Anadara multicostata* and *Megapitaria squalida*) are sand dwellers. Not all

TABLE 2. Pleistocene gastropods (phylum Mollusca) from the south side of Isla Coronados

Species	Dominance	Species	Dominance
Alabina strongni	R	*Oliva davisae*	C
Bulla aspersa	R	*Phyllonotus princeps*	R
Cassis coarctata	R	*Polinices bifasciatus*	R
Crepidula onyx	R	*Strombus galeatus*	C
Conus bartschi	R	*Strombus granulatus*	R
Conus recurvus	R	*Terebra variegata*	C
Conus regalitalus	R	*Trivia solandri*	R
Fasciolaria princeps	C	*Turbo fluctuosus*	R
Modulus cerodes	R	*Turbo squamiger*	R
Murex elenensis	R	*Turritella gonostoma*	R

NOTE: With respect to rank dominance, A = abundant, C = common, R = rare.

were alive at the same time. After all, the embankment is relatively thick. What the deposit confirms with certitude is that the Pleistocene shelf on the south side of Isla Coronados was relatively shallow and enjoyed mostly calm conditions.

The air temperature remains pleasant enough, but the sun is bright. Our tabulations have enjoyed morning shade beneath the embankment at the side of the arroyo. It is time to move on, but the way ahead will be under full sun. For the next tenth of a mile (160 m), the ground is open along the talus below the limestone bluff. Thereafter, more serious vegetation hampers our progress. After pushing through a tangle of brush, the crew arrives at an open plain covered with fine white sediment. The ground is soft and boots initially sink to the ankles as we cross between the shore on our right and the bluff on our left. Here, the shore is formed by a berm of smooth cobbles and boulders composed of andesite and basalt. This means that the fine, white sediment is not part of a beach deposit that came from the sea. Understanding how it may have been moved seaward from the adjoining bluffs is a puzzle that begs to be solved.

The white sediment thickens to the east and becomes firmer underfoot. Picking up a handful of the sediment, I ask one of the crew to examine it closely with the hand lens. Many of the larger particles show the smooth, bud-like ends of fruticose red algae from fragmented rhodoliths. When I feign utter confusion, I am reenacting my first encounter with this segment of the shore from January 2004. Where did this stuff come from? Yes, the algal origins of the bioclasts are easy enough to recognize, but how did the degraded rhodolith material arrive here? We are not on a beach. Another change in the landscape may have been overlooked. The low bluffs on our left are no longer formed by limestone. They have the reddish look of rusted andesite. Once realized, this information makes things even more puzzling, because pulverized rhodolith debris cannot be extracted from a volcanic rock. It is a mystery that becomes resolved only slowly with our onward progress eastward. Sharp eyes detect an opening in the andesite ridge, which we change directions to enter by marching north (figure 27).

Figure 27. Line of students entering the narrow opening to a Pleistocene lagoon below the quiescent volcano on Isla Coronados. Photo by author.

At first, it is like a cleft in the ridge that barely opens wide enough to admit sight of the island's hulking volcano. Once we are through the defile, however, a plaza emerges blazing white in the sun. We have reached Cañada Coronados (Johnson et al. 2007), a shallow basin more than 1,900 square yards (1,600 m^2) in size, carved from poorly consolidated limestone composed of medium- to coarse-grained calcarenite almost exclusively made of crushed rhodolith debris. It is not quite like entering the great Zócalo in Mexico City, but there is a sense of wonderment to be had in arriving unprepared for the enormity of such a space open to the sky. The rhodolith sand was originally impounded behind a barrier formed by the andesite ridge.

Excavation of the shallow basin is the result of relatively recent and ongoing erosion due to infrequent rainstorms. The water ponds, but finds an outlet through the opening in the ridge. This is the same opening through which we have just entered. When it rains hard enough, the water erupts from the opening and carries with it a load of white sediment scoured from the basin's sides and floor. In effect, the white deposit outside the barrier is akin to a microdelta. It is thickest outside the hole in the andesite ridge and includes the larger bioclasts of recognizable rhodolith fragments. A finer rock powder is carried off to the fringes of the delta, where we first encountered it.

More astonishing, there are fossil corals (*Porites panamensis*) preserved in growth position as thickets attached directly to the walls and upper slopes of the andesite ridge. Some colonies adorn the walls of the passage through the ridge. It means that the opening is not only a contemporary physical feature but also existed during the Late Pleistocene, when the place was awash with seawater and corals had the opportunity to occupy the surface. What we have under foot is a "fossil" lagoon that formed below a quiescent volcano with a set of barrier islets made of solid andesite marking the seaward margin.

The intricacies of the paleolagoon make a dense tapestry, the thread of which would be difficult to discern without the ongoing excavation by erosion of Cañada Coronados. The basin is deepest where we stand inside the entrance but gradually shallows in two directions, to the northeast and northwest, as separated by an intervening divide of strata. There is a kind of sensory overload, and it is hard to know where our attentions should be trained. To start, the plaza floor is strewn with the shells of fossil mollusks washed out from nearby canyon walls. Long and elegant, some nearly complete specimens of the pen shell (*Pinna corteziana*) are present where Durham (1950) first found them. Many of the same bivalves and gastropods tallied earlier outside the lagoon also are here. Among bivalves not noticed previously are the wedge-shaped shells of *Arca pacifica* and elongated "jack-knife" shells of *Tagelus californianus*. Among fossil gastropods not seen previously are the common cup-and-saucer shell (*Crucibulum imbricatum*), as well as rare specimens of *Malea ringens*, *Vasum caestus*, and the intricate sundial shell (*Architectonica nobilis*).

Wading farther into the Pleistocene lagoon, we cross a basin floor composed of coarse rhodolith sand now cemented into a solid pavement. Because we know that living rhodoliths are accustomed to some degree of free movement on the sea floor, a troublesome question resurfaces: how did such a voluminous mass of rhodolith sand get here? Although the paleolagoon is sizable, it is sheltered from the north winds by the volcano. The solution to this conundrum stares back at us from the east wall of Cañada Coronados. Rhodolith sand is preserved in distinct layers exposed by erosion in a vertical plane perpendicular to the orientation of the andesite islets fronting the paleolagoon. It is striking that the sandy layers are not horizontal but dip north into the basin off the top of the andesite ridge at an angle of 20°. Hence, the rhodolith sand was not produced inside the lagoon but imported from outside as a succession of washover deposits.

Similar events occur today during major storms, when seaward sediments wash across low barrier islands to become buried on the landward side in adjacent back-island channels. With a cumulative thickness of 21.33 feet (6.5 m), successive layers of rhodolith sand vary from 13 to 19 inches (40–60 cm) thick. Each layer represents another washover event related to a storm that swept northward through the Carmen Passage. This is the interpretation by Ledesma-Vázquez and colleagues (2007) for the sedimentary process that filled the Coronados paleolagoon during its early phase of development. In contrast to the south-dipping ramp predicted by me before reaching the island, the bare facts of the matter speak eloquently of a different scenario. Generated as subtropical depressions, the storms were major events that ran up the Gulf of California from the south. Although such storms are relatively rare in frequency, the long arc of time was adequate for a succession of them to leave a thick deposit inside the paleolagoon.

Storm deposits are not well recorded on the south shores of the island outside the lagoon. It must be assumed that diverse and well-preserved fossil assemblages (tables 1 and 2) accumulated and became buried there during intervals of calm. Once washed into the Coronados paleolagoon, the carbonate sand remained loose. As we walk over the now-well-consolidated floor of the basin, evidence as to its original nature is demonstrated by the many articulated shells of *Codakia distinguenda* preserved in growth position within the sand. These relatively large bivalves, up to 4.5 inches (14 cm) across, were among the early successful inhabitants to benefit from conditions in the lagoon.

Moving inland, we follow the canyon's northwest branch because it is well entrenched and provides better exposure. About 330 feet (100 m) from its small opening, the plaza-like character of the place takes on dimensions more like a true canyon, with opposing walls 200 feet (61 m) apart. Here we witness the most extraordinary part of the paleolagoon, and surely the island's great prize. It is a coral reef (figure 28). Erosion has brought large blocks of coral tumbling down the sides of the canyon walls. Individual coral heads have rolled nearly to the canyon's center aisle. A continuous line of coral colonies remains in original growth position along each canyon wall. These corals (*Porites panamensis*) are large, up to 35 inches (110 cm) in height. The sensation of walking among them is like exploring a coral garden. With long, armlike branches expanding from a central point where the colony first began to grow, their form looks ever so much like a huge bouquet of flowers. It is an appropriate comparison, because the class of corals to which these fossils belong is known as Anthozoa, or "flower-animals."

Unlike *Codakia distinguenda* and other bivalves of its habit such as *Tagelus californianus* that sought refuge within the rhodolith sand, coral colonies required a firm spot on which to anchor and grow above the sediment surface. Close examination of individual coral heads still in place on the canyon walls reveals that colonization of the lagoon by corals required stabilization of the rhodolith sand. Most of the colonies are attached to andesite boulders (see figure 28, inset). A few are fixed to large gastropod shells belonging to *Strombus galeatus*. Such a houseguest would be an unwelcome burden to the gastropod. The sea snails were no longer alive when commandeered as pylons.

Figure 28. Outcrop showing fossil corals (*Porites panamensis*) preserved in growth position (Ramón Andrés López-Pérez seated for scale). Inset shows an immature coral colony attached to an andesite cobble. Photos by author.

With upper branches spread equal in radius to overall height, the bouquet shape of the biggest corals spelled doom for lesser colonies caught beneath in the understory. The reef deposit includes immature corals (see figure 28, inset) that lived perhaps only a few years before losing out in the competition for space. In general, each mature colony from this part of the reef acquired about 10 square feet ($< 1\ m^2$) for living space. *Porites panamensis* is an extant species. A recorded growth rate for colonies living in warm tropical waters off the Pacific shores of Panama is less than 0.2 inches (5 mm) per year.[4] If the same rate applied to the largest fossil corals in the Coronados lagoon, their life span was long, indeed, although still in the realm of ecological time as opposed to geological time. Another question altogether regards the geological age of the reef deposit. A laboratory analysis for uranium/thorium ratios calculated from a sample coral (Johnson et al. 2007) illustrates that the reef lived about 121,000 years ago during the Late Pleistocene. Overall dimensions of the reef are difficult to appraise, since our exploration is limited by the confining walls of Cañada Coronados. In this part of the canyon alone, which entails an excavated space of about 4,785 square yards (4,000 m^2) enclosed on three sides by walls with intact coral, the number of colonies destroyed by erosion can be estimated as approximately 4,000.

Correlation from one canyon wall to its opposite number confirms that a pavement of andesite cobbles and small boulders separates the reef from the rhodolith sand below. Thus, we must ask how such a cover came to be introduced. Yet again, the advice of Charles Lyell is appropriate: seek clues in the modern environment. In turning to look back over the path we have taken across the canyon floor from the paleolagoon's entrance, we can see that a surplus of dark andesite pebbles and cobbles peppers the otherwise white surface. The same cloudburst that eroded rhodolith sand from the lithified overwash deposit in the recent past also brought bits of andesite into the canyon from slopes above. These loose clasts, too heavy to be flushed from the basin, remain from the last floods. Much the same process occurred as a distinct event under sufficient intensity to lay down a sheet of larger andesite boulders during a heavy downpour in later Pleistocene time. Had this prehistoric event never happened, coral larvae would have failed to find a secure place to settle and mature as colonies that we now find as fossils.

For however long the reef prospered, the same andesite pavement briefly hosted a diverse fauna that took advantage of substrate stabilization. Among the fossil bivalves left in growth position cemented to andesite cobbles are the jewel-box shells *Pseudochama janus* and *Chama mexicana*. Other articulated bivalves associated with this layer include *Modiolus capax*, *Barbatia reeveana*, and *Arca pacifica*, all in the habit of attaching to rocks with their wirelike byssal threads. Known from living counterparts to favor rocky shores, fossil gastropods such as *Turbo fluctuosus*, *Nerita bernhardi*, and *Acanthina tuberculata* also can be spotted among the cobbles that armor the surface. In contrast, scarcely any marine invertebrates are found to have inhabited spaces among the corals. Except for long worm tubes cemented at the sides of coral branches, there is little evidence of a diverse reef community.

A notable feature of the corals is that they gradually decrease in stature as we approach the back margin of the reef. Also attached to andesite boulders, the last colonies visible in the canyon walls stand hardly more than eight inches (20 cm) high. The change was due to the gently sloping surface on which the andesite clasts were entrained. Here, the highest reach of the coral branches is even with those colonies downslope toward the front of the reef. As with many other branching corals today, the reef can be assumed to have grown upward close to the level of low tide within the paleolagoon. Finally, we see that a thin layer of carbonate detritus largely devoid of rhodolith sand buries the coral reef.

Having traversed the length of the northeast canyon, we step onto a low incline—almost a plain. It stretches to the foot of the volcano a half mile (800 m) away (see map 8). There is no trail, and we must weave our way among the thorny palo adán trees and other bushes. From a geological perspective, the most telling feature of the surface is that it consists of white carbonate debris with the scattered shells of small bivalves and gastropods attesting to a former marine environment. Upon reaching an elevation about 115 feet (35 m) above sea level, we intersect the official trail that leads to the island's summit. With the noon hour approaching, we hasten westward to the beach where our hike began and our lunch supplies await.

Though discontinuous, the trail is marked by a double row of andesite cobbles outlining the margins. Soon enough, we find traces of coral heads where the pathway descends through what remains of another fossil reef. Now the trail dips below into the white rhodolith sandstone where the articulated shells of *Codakia distinguenda* are again plentiful. Just as abruptly, the trail crosses a gulley and mounts a low ridge formed by coral limestone. As we continue southwest, a grand view opens over the wide west bay with its rim of white sand. Where the trail abruptly descends the ridge, a succession of coral layers is well exposed. Each layer consists of colonies smaller than those observed at Cañada Coronados but likewise preserved in growth position. Given that we are at the same elevation as the main reef structure in Cañada Coronados, these corals must be about the same geologic age. The repetitious aspect of coral layers in the ridge hints that the local environment was prone to disturbances not felt in the more sheltered paleolagoon with its protective front of andesite islets.

Time travel through the Gulf of California is a hot and dusty business. The still waters of the bay have warmed during the morning hours, and they invite a refreshing swim before lunch. It is fitting to leave the drained lagoons of ancient landscapes aside for a while and rejoin the present to splash about in actual water. The reality of a hungry stomach interrupts before too long, however, and demands access to the food chests. The palapa provides welcome shade from the midday sun, but the fearless take to the open beach and soon dry off. We are fed, rested, and ready for the second hike of the day by 1:30 p.m. Our trek will bring us to the summit of the volcano.

When we have returned to the trail along the coral ridge behind the beach, it is instructive to look north across the gulley to the rugged field of red volcanic rocks covering the southeast quadrant of the cone. The dry streambed below us is white with limestone and related debris, which shows us the location of a former paleoshore. Above this contact rises a surface with a slope of 8°, the most gentle anywhere around the lower slopes of the extinct volcano (see map 8). Based on the mapping of Bigioggero and colleagues (1988), this quadrant represents the volcano's youngest lava flow, dated to 160,000 ± 20,000 years ago. The magma was not the fluid sort that spills down an incline to eventually solidify as ropy basalt (pahoehoe lava). What we see is more akin to the blocky, jagged deposit of a slow, creeping flow the Hawaiians called "aa" lava. Be that as it may, the least incline on the Coronados cone is not the place to initiate a climb. Such a route would be torturous, ending in shredded boots before much of any height was gained.

The marked path brings us almost midway across the island, before turning northeast to start the ascent over an older part of the volcano, dated to 690,000 ± 50,000 years ago by Bigioggero and colleagues (1988). Here, the path is hardly more than a footstep in width rising steeply through a field of angular boulders (figure 29). Reaching the 330-foot elevation (100-m contour), the trail levels off for nearly a third of a mile (500 m) along a southeast-directed ridgeline (see map 8). It is a strategic place to pause, catch one's breath, and look out over the apron of Pleistocene limestone attached around the waist of Isla Coronados. Looking rather diminished in the distance is the island's Zócalo, a two-pronged white gash in the landscape. Adjacent to the white plaza on its eastern side is the rise of land that first caught my attention when viewed in profile from Punta el Bajo on the peninsular mainland. From here, the southeast corner of the island looks

Figure 29. Ascent of the Pleistocene volcano on Isla Coronados. Photo by author.

more like a small sombrero than the ramp-like structure I mistakenly projected. This is the oldest part of the island, with Miocene andesite attributed to the Comondú Group by Anderson (1950). We contend that the andesite islets forming the outer barrier to the Pleistocene lagoon are Comondú equivalents that further underlie the island's great west arm (see map 8). Twelve million years old (based on dates from our own samples), these rocks are the island's foundation.

Not easily discernible on the flats below, remnant terraces situated 52 feet (16 m) and 82 feet (25 m) above sea level include scattered fragments of coral fossils (López-Pérez 2012). They occur above the plaza reef toward the margin of the "sombrero." The implication is that a longer Pleistocene history of colonization by corals is recorded on the sheltered south side of Coronados. As yet, no radiometric dates are available from these corals, but should the terrace interpretation hold valid, it means that Isla Coronados was subjected to episodic uplift sometime after its volcanic eruptions 690,000 years ago. The most recent episode postdates the plaza reef itself, the base of which now resides close to 39 feet (12 m) above sea level at the rear of Cañada Coronados. Perhaps only half this elevation is due to a higher stand in sea level at the end of the last interglacial period 121,000 years ago, while the rest is attributed to local tectonics.[5]

Now that our heartbeats have returned to normal after the first sustained push against the summit of Coronados, the second phase of the climb awaits. Leaving our lookout on the ridge, we are confronted by a slope of 33° over loose material eroded from the top part of the volcano. The last segment of the ascent is routed straight up the volcano through more than 460 feet (140 m). It is slow going, and I take the place of honor bringing up the rear. Once on top at a height of more than 850 feet (260 m) above sea level, we leave the shelter of the south face and are greeted by a northerly breeze. The trail crosses a shallow saddle and continues a short distance to the north to provide a crowning view. The steepest slopes on the island are located on its eastern and northern flanks. These support some of the oldest flows from the volcano, dated by Bigioggero and colleagues (1988) as approximately 1.8 million years old.

From our perch, the view is uninterrupted to El Mangle, situated 12.5 miles (20 km) to the northwest. As described in chapter 6, the entire arc of limestone strata can be seen along the coastline of the North Loreto basin (see map 6). Punta el Mangle, itself, is clearly visible. The delineation of El Coloradito Fault springs forcibly to mind, but Isla Coronados does not align itself on that potential fracture of a transform fault. Why did the Coronados volcano erupt here, and not somewhere else? There is no ready answer. Perhaps a crustal weakness related to the older flows of equivalent Comondú lavas had something to do with the plumbing of Pleistocene magma to build this impressive edifice.

The breeze intensifies. Due north over the empty surface of the dark-blue Sea of Cortez, the horizon has become hazy. It is a sure sign that a stiff wind is on its way south, blowing up a fine mist of salt aerosol. The aerosol is responsible for deforming the small copal trees (*Bursea hindsinia*) at the summit into classic *krummholz* (German for twisted wood). It is time to beat a retreat to the beach. While the measured march up the summit trail required two and a half hours (with ample time for a pause), the downhill return is covered in 45 minutes. The boat captains are waiting for us on the beach when we arrive 15 minutes before scheduled departure. The wind is fully at our back on the return trip to Loreto, and the boats ride the swells with added vigor to the push of the engines.

Loreto Bay National Park (Bahía de Loreto Parque Nacional) was created by decree through the office of the president of Mexico on July 19, 1996. It was a bold move that conferred conservation status on the several islands and marine territory within the nearly 775-square-mile (2,000-km^2) area of the reservation. Located in the parking lot near the boat ramp on the Loreto harbor is a bronze plaque that commemorates the addition of the park to the list of World Heritage Sites sanctioned by the United Nations Educational Scientific and Cultural Organization (UNESCO) in 2005. The UNESCO certification puts into succinct language the park's "outstanding universal value to humanity" as a world-class eco-region worthy for its natural heritage. Exceptional geological features also add distinction to the park's natural monuments reflective of the power of winds, waves, and other physical processes that shaped biological geography through the last several million years in the Gulf of California. Most telling of Isla Coronados is its battered northern shore, its western channel with the ideal nursery for rhodolith beds integral to carbonate beach and dune development, and its sheltered south shore with a history of coral reefs. Coronados is that ideal island through which a refined understanding of the natural economy at work throughout the Gulf of California may be understood.

While the island is a key focal point for the education of its many human visitors, it remains a place fully wired to the ongoing natural rhythms of the gulf. Other visitors come to Coronados on a regular and life-sustaining basis. In January 2006, Dave and I, together with two students, arrived on Isla Coronados with the mission of visiting a small pocket beach located a little more than a half mile (900 m) northeast of the main beach on the west side of the island. The pocket beach intrigued us as another spot where carbonate sand was actively accumulating, but it was painfully difficult to reach by walking through the young lava field that reaches down to the shore from the southwest slope of the island. Once our boatman landed us on the beach, we realized that we had chanced upon the death throes of a group of Humboldt squid (*Dosidicus gigas*) intent on reproduction in the final phase of their life cycle.

There were a score of the jumbo squids, bright red in color, both males and females, caught in the act of releasing eggs and spawn into the shallow water. Individuals were left exposed by the receding tide (see plate 11). Spent of their biological imperative, the large invertebrates made no attempt to exit the small bay and return to deeper waters. The gulls were onto this rare booty from the sea and had begun to alight on the beach. Though cautious of our presence, they attacked the still-writhing squids on our periphery with gusto—showing particular interest in the eyes and general head region. Our boatman also sprang into action, choosing one of the larger specimens to cut into strips of flesh to use as fish bait. We were forcibly awestruck by the great wheel of life and the impeccable way that all life is recycled in the Gulf of California. Searching the dry sand higher on the beach, we found ample evidence—by way of the sharp, horny beaks typical of squids—that the enduring life cycle of the species had a previous history on this particular beach. It is but one of the many lessons I am reminded of on departure from the Loreto airport when the flight path crosses the red volcanic island with its luminous turquoise waters and fine, white beaches (see plate 12).

8

Song of the Amazon on Isla Monserrat

To me, the clerk's account of the giant women and their early paradise had seemed as fictional as Montalvo's imaginative novel . . . , in which he told of the goddess-like Amazons. That is, it had before an expedition during which I got a panoramic view of the region.

—*Ray Cannon,* The Sea of Cortez

THOSE WHO PERUSE MAPS as a pastime know that the name *Monserrat* has regional spellings as variable as *Montserrat* in the Catalonian district of Spain and *Monceaux* in French Normandy. As a geographic moniker, the word takes its derivation from the Latin *mons serratus,* for jagged mountain. Two islands have adopted the name. The larger and better known is Montserrat, the British overseas territory in the Lesser Antilles. It is infamous for its Soufrière Hills volcano, which after centuries of dormancy roared to life in 1995 with major eruptions following in 1997, 2009, and 2010. The volcano's devastation forced an exodus of roughly half the island's 12,000 inhabitants. In contrast, Isla Monserrat in the Gulf of California is a much smaller island (7.15 sq. mi. or 18.5 km^2) having extensive volcanic rocks but no history of settlement. The Mexican island is part of Loreto Bay National Park and enjoys full protection under park regulations. Requiring landing permits, the island is seldom visited even by the cruise ships that specialize in nature tours.

The direct route from Loreto southeast to Isla Monserrat covers 28 miles (45 km) over exposed waters for the most part. Landings at the north end are seldom made, owing to the strong sea swell that impacts the beach. Secure anchorages are found at the south end of the island. With all necessary permits in hand, I have made the trip twice by open boat with colleagues and students. The first visit, in January 2003, went smoothly, and I questioned the boatman's adamant refusal to discharge us at the north end. Getting onto the beach under calm conditions was no problem for a stay of a few hours, but retrieving our crew after a visit of several days could prove hazardous—I was told. My second excursion, in January 2006, was under rough conditions that gave me a better appreciation for the situation. My windbreaker failed to keep me dry during that trip. Disembarkation was on a cobble beach at the leeward end of the island (map 9). On both occasions, we happily made camp on the low bluffs, above the beach.

Isla Coronados may be the most perfect island in the Gulf of California (chapter 7), but Isla Monserrat is the most mysterious. Ray Cannon (1966, 146) coined a geographic name for the isolated region around Monserrat that includes the peninsular mainland at Agua Verde and Isla Santa Catalina (see map 1, locality 8). He called the place Juanaloa, meaning "praises to Juana." However tenuous, the existence of such a woman, known as Juana (or Jane), has some claim to fact, rooted in a visit to the La Paz area by none other than Hernán Cortés in 1535. A clerk to Cortés purportedly described an Indian woman of great stature as well as "elegance of face and form" who had been kidnapped from the "domain of the Amazons" farther north. Cannon adopts the Spanish phrase "*reino de felicidad ultimo*" (realm of ultimate happiness) to characterize "Jane's" homeland. Said to be isolated by impassable mountains, the inhabitants lived in an earthly paradise that afforded abundant fruit, wild honey, game, and seafood fished from the bays and shelves of nearby islands. It is a tale of the kind only a teller of fish stories such as Cannon would embellish. Having explored all corners of the Sea of Cortez, the intrepid fisherman was awed by the beauty and tranquility of a place he attributed to the home of the Amazons foretold in the fictional account of Montalvo.[1] The remoteness of Juanaloa has much to do with its inherent mystery. Little is written about the region. The road to La Paz

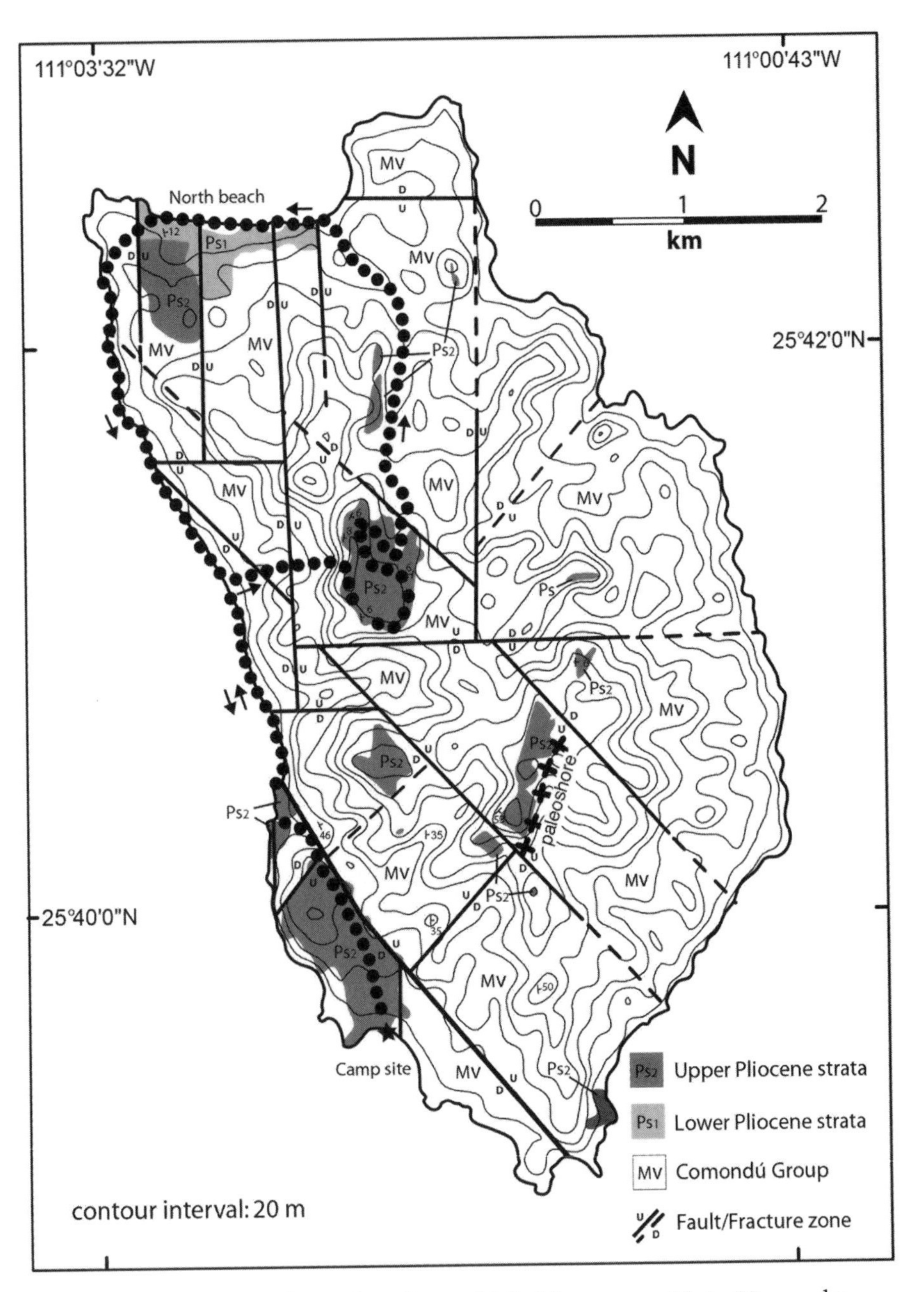

Map 9. Topography and geology of Isla Monserrat. *Note:* Heavy dotted line (with directional arrows) shows path of featured hike; heavy black lines represent dissecting faults with relative up (U) and down (D) movements so labeled. Conventional strike and dip symbols relate the orientation of sedimentary layers. Map by author.

makes a wide detour inland below Loreto (see map 1) to circumvent the coastal mountains. One can easily miss the turnoff to the village of Agua Verde, from which Isla Monserrat is visible. The difficulties of penetrating Juanaloa overland are heightened by the propensity of the inhabitants to remove the way-sign for the Agua Verde road.

It was the promise of the island's geology and paleontology that brought me to Isla Monserrat. Scarcely more than a single page of notes on the island is available from Anderson (1950), who begins with an apology for having devoted only part of one day to its exploration during the 1940 cruise of the research vessel *E. W. Scripps*. Noting that high surf made a landing at the north end impossible, the geologist went on to explain that his limited time on the island was restricted to the south end. Anderson must have been in excellent physical shape, something we grew to appreciate in our attempt to follow his tracks during our first foray on Monserrat. As we coasted the west side of the island before reaching camp, the geological feature that made the strongest impression was the lofty position of Pliocene limestone seated on high cliffs above Miocene andesite. The color contrast, white on red, is the same as that found many other places on the gulf coast of the Baja California peninsula. Offset by a succession of faults looking like giant steps, it is the elevation of the topmost limestone step that stands out as extraordinary.

From his description of the "angular discordance" between tilted andesite layers and the basal conglomerate of eroded andesite clasts beneath the limestone, it became clear that Anderson was an able climber. We needed the better part of a day to confirm what he had achieved. On the whole, Monserrat is not a particularly high island. Neighboring Santa Catalina, for example, rises to an elevation of 1,542 feet (470 m), while Monserrat attains only 734 feet (224 m). The difference is that Catalina has no limestone, while patches of limestone layers left intact on Monserrat are like a tattered tablecloth covering elevations exceeding those of Pliocene limestone elsewhere in the gulf. As a reminder, Pliocene limestone at El Mangle (chapter 6) begins at the water's edge but rises 130 feet above sea level (40-m contour) in the landward direction. The inner margin of the Pliocene basin at San Francisquito (chapter 4) sits 480 feet (146 m) above sea level. In both examples, the conglomerate between basement rock and overlying limestone is a reliable reference point to former coastal

Figure 30. Fossil echinoids (*Clypeaster bowersi*): *left,* aboral view of complete test; *right,* oral view of abraded test. Photo by author.

conditions at or near sea level. Differences in altitude where the conglomerate and related limestone strata occur today are due to a combination of sea-level changes and uplift related to regional tectonics.

A short distance west along the shore from the campsite on Monserrat, the limestone reveals fossils belonging to the sea urchin *Clypeaster bowersi* (figure 30). Armored like a Sherman tank, the tests of this robust echinoid are large, up to six inches (15.24 cm) in diameter. The extinct species occurs in layers older than those with *C. marquerensis* found at El Mangle. Thus, strata on Monserrat were deposited somewhat earlier during the Pliocene than was the ramp sequence at El Mangle (chapter 6). As noted previously, the biostratigraphic relationship of such index fossils was worked out inductively from field observations made by Durham (1950), the paleontologist on the 1940 cruise of the research vessel *E. W. Scripps.*

Following an arroyo north along the east fault scarp of the limestone near camp during our first visit, we pushed inland a short distance to a tilted block of Miocene Comondú that rises like a monolith blocking the way. Thick layers of the red andesite incline due east at

Figure 31. Rotated fault block of Miocene Comondú rocks with volcanic layers dipping to the east. Photo by author.

an angle of 35° (figure 31). It is a pattern consistent over much of the island. Long before the Gulf of California opened, much of the entire region was buried under layer after layer of volcanic flows consisting of andesite and coarse andesitic agglomerates, as well as tuff from ash falls. Geologist C. Carew McFall (1968) was the first to formally subdivide and appraise just how thick these various deposits might be around Bahía Concepción to the north (see chapter 5). The tilting of fault blocks in the Comondú occurred as a consequence of east-west stretching and crustal failure.

At Punta Chivato north of Mulegé (see map 1), fault blocks of comparable size dip at an angle similar to those on Monserrat but incline uniformly to the west (Johnson 2002). Comondú fault blocks throughout Baja California rarely exhibit dip angles in any direction other than east or west. During the extensional phase of regional tectonics, relatively flat plains resting on a thick pile of volcanic layers were converted to north-south ridges defined by north-south fractures. Secondary fractures released strain that resulted in rotational slumps of smaller crustal blocks directed one way or the other normal to the fractures. On a larger scale, primary fractures promoted the development of structural grabens that formed sizable rift valleys separated by structural horsts. The action began some 13 million years ago during the middle Miocene and continued into the early Pliocene, when the lower Gulf of California was flooded by ocean water (Ledesma-Vázquez et al. 2009). The Carmen Passage between Isla del Carmen and the peninsular mainland (see chapter 7) is one such topographic adjustment related to a graben. The same is true for the channel between Monserrat and the peninsular mainland around Agua Verde. Essentially, Monserrat is a structural horst transformed into an island during the flooding of the gulf that experienced uplift long after the first limestone deposits accumulated.

The conglomerate described by Anderson (1950) as 10 feet (3 m) thick is nowhere in the vicinity of the campsite. During our 2003 visit, it became a priority to find where he observed those rocks. Our mission to map the paleoshore entailed a full-day exercise that led me on the second most punishing hike I have endured in Baja California. To reach the nearest trace of conglomerate sandwiched between andesite (below) and limestone (above), we were obliged to find our way around the obstructing Comondú fault block (see figure 31) and up the side of a steep ridge through nearly 330 feet (100 m) of elevation. There, a patch of limestone drapes a notch in the ridge eight-tenths of a mile (1.3 km) northeast of the campsite at an elevation of 575 feet (175 m). The associated conglomerate is only 1.6 feet (0.5 m) thick, with andesite cobbles that rarely touch one another. Anderson had gone farther.

The next rise in topography is offset by a northwest-southeast-trending fault that brings the top of the limestone to an elevation exceeding 655 feet (200 m) above sea level (see map 9). A hard scramble brings the explorer 525 feet (160 m) higher than the ramp top at El Mangle or 177 feet (54 m) higher than at San Francisquito. Here, the limestone cap is underlain by a suitably thick conglomerate of cobbles and boulders as large as 16 inches (40 cm) in diameter. Fossils entombed among the andesite clasts include the same sea urchins (*C. bowersi*) as those observed near the campsite, augmented by a diverse fauna of mollusks dominated by articulated *Glycymeris maculata*. Less commonly, the conglomerate includes fanlike molds of pecten shells. Steeply inclined volcanic layers worthy of Anderson's "angular discordance" support the conglomerate. We had little doubt that our predecessor stopped here. Additional biostratigraphic insight from the limestone cap at this locality is shown by the abundant pecten shell *Argopecten antonitaensis* in a matrix of rhodolith particles. The pecten is another good index fossil that overlaps with the upper range of the sea urchin (*C. bowersi*). It means that the timing of the limestone onlap is restricted in age to the upper (later) range of the sea urchin. The matrix also tells us that a rhodolith bank was nearby.

We were inspired to follow the limestone ledge as far inland as it might take us. Scarcely another quarter of a mile (400 m) to the west brought us to a zone where the limestone thinned to a featheredge and andesite bedrock emerged from below to trace out a paleoshore (see map 9). Amid fossil corals (*Favia maitreyiae*) weathered from the thin crust of limestone (López-Pérez 2012, 38) are abundant gastropods including the oversized strombid (*Strombus galeatus*) and the smaller, distinctively ribbed neritid (*Nerita scabricosta*). None of these fossils were reported previously from Monserrat, which gave us confidence that we stood where no other geologist had come before. The fossil neritids made an indelible impression, identical to the extant species so strikingly photographed by Herman Zwinger from a tide pool in the southern cape region of Baja California.[2] The Monserrat paleoshore can be followed nearly a half mile (750 m). Walking the ancient shore felt just as real as a stroll along the modern rocky coast near the dramatic tip of the Baja California peninsula. The essential difference, of course, was the altitude of our path, confirmed by ground-positioning system (GPS) as 670 feet (204 m) above the present sea level.

On that initial hike into the southern uplands of Monserrat, I was determined to return to camp by a more circuitous route. Other patches of surviving limestone to the north stood high on pediments of dissected andesite. I thought we should be able to reach one of them by following a bearing more or less west along a prominent ridge (see map 9). In the end, it was a strenuous trek that brought members of the team to the point of exhaustion on reaching the island's west coast about three-quarters of a mile (1.2 km) north of the campsite. Bone-weary, Dave and I staggered into camp at the end of the day and sent the boat to collect the others waiting on the shore. It was on that excursion, however, that I recorded the landscape photo of which I am most proud, above all others from the hinterlands of Baja California. Even though my feet were firmly planted on the ground where the photo was taken, it gives a bird's-eye view of a great limestone plateau seated high on the broken terrain of andesite layers tilted steeply to the east (see plate 13). The moment the photo was taken, I realized that reaching the plateau that same day would be impossible. Our 2003 trip was broadly exploratory in nature with no plan to stay very long in one place. We had visited Isla del Carmen for the first time, and now Monserrat. A long boat trip to Isla San José to the south remained before us. I knew that we would have to return to Monserrat on another occasion.

Once we were back on Monserrat in 2006 with a new team, geology student Elizabeth Pierce (a star field-hockey player in top physical shape) protested that Dave and I were doing our best to kill her during our climb to revisit the Anderson locality. Meanwhile, Jorge and his graduate student Astrid Montiel-Boehringer busied themselves near the campsite collecting data on the Pliocene limestone rich in fossil pectens. The last day of fieldwork on the island was reserved for my assault on the high limestone plateau in the central part of the island. Dave and I would go alone. It turned out to be the single most physically demanding but mentally elating hike of my career in Baja California. The daylong trip outlined on map 9 is based on our ten-mile (16-km) trek of January 20, 2006, which brought us to the north end of the island before our return to camp at the close of the day. No one should attempt this hike alone. Permission from authorities at Loreto Bay National Park is mandatory.

FEATURE EVENT

Siren Call of the Last Amazon

Night falls in utter blackness interrupted by a late moonrise. Sleep comes fitfully with unease over the coming day's excursion to the center of the island. Dreamland is a place in a parallel world, made oddly recognizable by flashes of cinematic reality from our wakeful hours but simultaneously bizarre with details that defy logic. The dream is mine alone. Garcia Montalvo and Ray Cannon cannot be held responsible, although their sway is undeniable. My tent is illuminated by soft moonlight. From outside, far away, issues the song of a woman in a strong contralto voice. Her song rises and falls with a delivery as honest as the cool night air. It is a funeral hymn, a requiem. The woman stands alone on the high limestone plateau, as if she has taken to the podium at the center of the world. Who does she sing for? Is it for the end of her kind, a people for whom the price of salvation was disease of epidemic proportion? Is her song for the intertwined ecosystems of the Gulf of California, threatened by ever-mounting human pressures? Is her song for me? I am fearful but powerless to prevent an out-of-body experience in which my consciousness floats up the steep side of the island to the rim of the plateau. As the last syllable of her song fades, she stands with arms outstretched in the pale moonlight. A sharp rock beneath my sleeping pad abruptly ends the dream, and I shift my body to another position in dumb slumber.

A sense of calm resolution pervades over breakfast. We will claw our way to the roof of the island, using the pick-end of our geology hammers if needed. With provisions for lunch and the bare minimum of equipment packed for the day's trip, we head off from camp to cross the limestone ramp north of camp. Three-quarters of a mile from camp (1.25 km), we reach the west coast, where it is possible to slide down to a wave-cut platform hugging the shore. Cut into andesitic agglomerate by scouring waves, the platform is roughly 18 feet (5.5 m) wide with a very gentle seaward tilt. Physical geographers would call it a Type A platform,[3] as differentiated from sea cliffs that plunge straight down through the water.

Here, the adjoining sea cliff is commonly more than 18 feet (5.5 m) high. Although the agglomerate shows evidence of cobble-size pieces that were plucked from the outcrop by erosion, the space where the platform meets the cliff face is remarkably clear of rock debris (figure 32). It shows that the waves during high tide are efficient in keeping the platform swept clean. The outgoing morning tide has not yet ebbed. A space about 6.5 feet (2 m) wide along the base of the cliff remains dry. In all likelihood, we will have a safe return

Figure 32. Sea cliff and wave-cut platform eroded in andesitic agglomerate on the west shore of Isla Monserrat (walking stick 3.6 ft. or 1.38 m in length). Photo by author.

passage below the sea cliffs even with the afternoon's rising tide. The platform is a superhighway that speeds us on our way north over the next eight-tenths of a mile (1.3 km).

Our opportunity to strike inland from the coast is under constant appraisal. The island rises steeply from the shore almost everywhere, and there are few good choices. A pause to examine the ridgeline high above suggests that we have arrived below a suitable topographic indentation on the island's flank. I remove my cap to mop my forehead; Dave shrugs his shoulders. The way inland will not be

Figure 33. Scree slope on west side of Isla Monserrat covered by veneer of broken limestone from the cap rock above. East coast of Isla del Carmen in the distance. Photo by author.

any better elsewhere (see map 9). The steepness of the first 50 feet (15 m) of the ascent is ridiculous, but thereafter the slope lessens and we are headed for a recess in the island's flank. Reaching an elevation nearly 200 feet above the water (60-m contour), we are surprised to find ourselves on a ridge that requires descent into a small valley before we can continue upward. It is discouraging, but we will not turn back. At an elevation of 330 feet (100-m contour), we arrive at the steepest part of the ascent and begin to pick our way upward over a surface of weathered andesite.

A patch of outcrop towers above, but it is fractured. Dave climbs first, testing handholds by which to leverage his body. His feet dislodge bits of andesite, and I instinctively crouch against the lower cliff face to avoid being struck. I follow after as best I can. Our progress brings us to a scree slope turned white by a thin cover of limestone debris fallen from the eroding cap rock above (figure 33). The angle of repose is 32°, but our goal is now well in sight. Leaning into the slope to keep a low center of gravity, we crab our way straight to the top side by side on all four limbs. Once we reach the unconformity, the last obstacle is to choose a way through the 10 feet (3 m) of limestone strata. Safely seated on the outcrop at the top, we pause to enjoy the clear view of Isla del Carmen some 7.5 miles (12 km) away on the horizon.

The first order of business is to acquire a satellite for a GPS fix. Here, the west rim of the plateau is 633 feet (193 m) above sea level. From below, the limestone layers appeared to be flat lying, but on the rim the strata are found to dip eastward. The clinometer registers a dip angle of 6°. Few macrofossils are apparent, which is a disappointment. The most obvious are scattered pecten shells (*Lyropecten subnodosus* and *Aequipecten abietis*). Neither provides a useful biostratigraphic resolution. More significant, the limestone is dominated by fine bioclasts derived from crushed rhodoliths. The broken bits look like white freckles in an otherwise gray-weathered rock. Eastward, sparse vegetation casts a dusty-green pallor over the plateau. Plants are evenly spread, dominated by dwarf copal (*Bursea hindsinia*) with fewer of the spidery palo adán (*Fouquieria diguetii*) in evidence. Some cardóns are present, but they are small. They have more the stature of a fence post than a mighty Amazon with outstretched arms.

We follow the west side of the plateau until gaining a position on the southwest rim (see map 9). The rim rocks here show a different orientation with beds that dip 6° northeast. Compass readings demonstrate that the limestone layers trace a delicate curve along the margin, inclining inward toward the center to take on the shape of a shallow bowl. In the distance, strata to the east appear to reflect a slightly upturned lip. We continue to the south end, where a breeze from the north catches hold. With the noon hour close by, we climb over the edge where the limestone rests on tilted andesite layers and pause to enjoy our meager lunch. The air is still. Seated just above the unconformity with our backs pressed against limestone strata, we gain comfort from only a thin arc of shade. The drop-off beneath us is unnerving. I do not fancy going down this slope to reach the bottom of the east-west valley 330 feet (100 m) below. The route presents itself as the most expedient way to start homeward later in the day. I inform Dave that I would rather endure any detour to avoid descending the plateau from this spot. My concern is met with thoughtful silence.

Before retaking the rim of the plateau, we look for fossils and break rocks to examine fresh surfaces for details of texture and grain size. The limestone here is much the same as that first encountered on the west rim. The basal conglomerate is just as poorly developed.

There is a sense of giddiness at having reached a place where few others would be so foolhardy as to venture. A plan of attack was safely executed to get us here, but now we must come home with the goods. What about the east rim? There, it is found that limestone layers dip 6° westward (see map 9). Indeed, the plateau is not truly a plateau at all. Strictly speaking, it appears to be a shallow basin. We are motivated to visit the northwest corner of the structure to further test this deduction. The most direct path is through the center of the basin, and we set out at once. With vegetation sparse, the going is easy.

Halfway across, limestone rubble is no longer underfoot. We are on a patch of ground covered by claylike soil, reddish orange in color. I suggest to Dave that the shape of the basin is a potential trap where fossils weathered whole from the limestone might be concentrated. A few steps farther, he stoops and picks up a complete rhodolith about an inch (2.5 cm) in diameter. Soon enough, we manage to collect a sample of diverse rhodoliths that reflect a range of morphological types and different genera.[4] The most abundant and largest (up to 1.5 in. or 4 cm in diameter) are delicate fruticose types. Nearly as large but not so common are lumpy forms with stout branches and stubby tips. Foliose rhodoliths with a bladed pattern are smaller and less common. Details characteristic of species from at least three different genera of coralline red algae (*Lithothamnion*, *Sprolithon*, and *Lithophyllum*) are represented in the assemblage (Johnson et al. 2009a). Because the slopes to the east, south, and west are gentle and no whole rhodoliths were observed in the surrounding rim rocks, it stands to reason that the assemblage was local in origin. In short, we traverse the center of a shallow seabed with a rhodolith garden that was thrust skyward on a rising island.

Advancing toward the northwest corner of the limestone cap, we find that the rim rocks are more elevated than elsewhere and dip at low angles between 6° and 8° inclined to the southeast (see map 9). Full closure into an actual basin is not, however, confirmed by limestone layers at the far north end. In point of fact, no south-dipping layers are in evidence. Isolated patches of Pliocene limestone on Monserrat, more of which can be viewed in the distance to the north, are like a secondhand jigsaw puzzle acquired at a yard sale. It is not new off the shelf, and many pieces are missing from the box. The challenge is to reinvent the lost pieces to complete the full image. Assuming that the surviving pieces retain their original relationships to one another with minimal lateral jostling on opposite sides of faults, one can concoct a plausible picture. For instance, we take it for granted that the semibasinal limestone "plateau" remains the same distance from the paleoshore to the southeast as it was during the middle part of the Pliocene more than three million years ago when the Monserrat horst was flooded. If so, the space between is reasonably restored as a connecting shelf that merges into the gentle depression on the sea floor reflected by our compass measurements.

Sizable indentations bite through the Pliocene limestone at the north end to expose Miocene andesite, which locally is softer than the cap rock. We are astonished to find our course blocked by a gulley 700 feet (213 m) long that rends the northeast quarter of the structure (see map 9). It is through this incision that any significant rainfall on the plateau is discharged as runoff. The present landscape makes a good deal of sense, because the limestone cap rock originally was deposited to conform to a modest, trough-like depression on the surface of the Miocene basement rocks. Thus, the former seascape exerted a profound influence on the development of the modern landscape. The nearest sliver of Pliocene limestone a third of a mile (500 m) to the north shares an alignment with the high west rim of the plateau. Likewise, the valley immediately east of that splinter shares an alignment with our gulley.

The north-facing gash through the Pliocene-Miocene unconformity gives me pause for thought. We ought to use the gulley as our escape route off the plateau. Of course, the northern route would entail a considerable detour on our homeward trek. The trade-off is an opportunity for exploration through the north-central part of the island. I must convince Dave that the detour is a superior plan, and I argue that the orientation of the gash presents a good angle for a safe descent. As we stand erect on our legs, we can easily control the rate of drop with each step. The relative softness of the andesite locally affected by geothermal activity also allows our boots to sink in and maintain stable footing. We are down the 150-foot (46-m) slope in a flash. The afternoon is coming up on 2 p.m., and we have a nearly two-mile (2.5-km) hike ahead of us to reach the north coast. The trick is to follow the most advantageous arroyo and keep any climbing to a minimum. Diverted too far from a steady north course, we could lengthen our hike to emerge on the wrong side of the island.

More so than any other spot I have visited in Baja California, I am feeling a sense of reverence for the Monserrat "plateau." There is no evidence at all that prehistoric peoples ever penetrated here, much less the lost members of a tribe of Amazonian women. Regardless, the place rises dramatically from the center of an island in the middle of a remote and achingly beautiful part of the gulf. With every step I take away from the structure, I exult in the simple joy that we were able to reach it and depart without mishap. We have caused no detriment to the setting, and we leave without doing bodily harm to ourselves. The dream song of the Amazon was not an omen of my premature demise. It was a conceit on my part to ever think so. Still, the plaintive melody echoes in my mind. I take it as a warning cry for preservation of the island's natural sanctity, bridging past and present.

On our gentle coastward trek, the sound of boots grinding against deep arroyo gravel accumulated over centuries is something never heard before in the narrow, twisting glens with their graceful palo blanco trees (*Lysiloma candida*) and occasional cardóns seemingly sprouted from bare rock. The basement andesite here is more competent than we have noticed elsewhere on the island. Dave calls a halt, now and again, to measure the inclination of layers that dip uniformly to the east and to record bearings on the zigzag course of our arroyo influenced, no doubt, by a raft of intersecting faults. It is 3 p.m. when we finally emerge on the island's shore at a rocky spine near the great north beach. Some effort is required to cross a low ridge of andesite separating us from the beach, but finally we stand at the east end of an expanse of sand more than three-quarters of a mile (1.25 km) in length (see map 9).

Due north from the beach, there is nothing but open water until the port city of Guaymas on the Mexican mainland 140 miles (225 km) away (see map 1). It means that the winter wind has a huge fetch over which to transfer energy to sea swells that eventually arrive on Monserrat. Partway along our traverse of the beach, I snap a photo of Dave, leaning on his walking staff fashioned from a cardón rib (figure 34). At least four sets of curling waves that make white water are entrained off the beach. They show the passage of waves over shallow sandbars on their way to the beach. Between us, the detritus left on the beach from the last high tide is peppered by dark pebbles of eroded andesite but otherwise well shingled by a layer of abraded shells. It is a scene that evokes in my memory the beach fronting Ensenada el Muerto on the north side of the Punta Chivato promontory (see map 1), a beach with a superabundance of smooth shell bits worn to the size of Mexican coins in the denominations of one and five pesos.[5]

Likewise, the sandy shelf bottom attached to the north side of Isla Monserrat is home to enormous populations of infaunal bivalves that live in safety from the waves below the water-sediment interface. Several mollusk species thrive under the import of well-oxygenated water infused with phytoplankton consumed via siphons that reach to the sediment surface. Only after the bivalves expire, at the end of a natural lifetime lasting some half-dozen years or more, are their

Figure 34. North beach on Isla Monserrat with extensive shell debris (Dave Backus for scale). Photo by author.

empty shells scoured out by the waves and washed shoreward. Disarticulated shells break into smaller and smaller pieces as they crash and grind against one another under frequent agitation by the waves. Over time, jagged edges of broken shells become worn and smooth. This is a special sort of sand beach that forms nowhere else on the shores of Isla Monserrat. It is a beach graveyard for generations of mollusks that lived and died in a territory all of their own as distinct as the living rhodolith banks off the west side of Isla Coronados (see chapter 7). But there is more to the Monserrat story. Just as the west-facing beach on Coronados supplies a dune dominated by finely milled rhodolith sand, the north beach on Monserrat feeds dunes that migrate inland with the filings of abraded mollusk shells blown off the shore. One such dune rises from the beach nearby, ramping upward to an elevation 72 feet (22 m) above sea level over a distance of almost 740 feet (225 m).

Time is limited, but I decide to pick up the next 50 or so relatively whole bivalve shells that I find on the beach. Soon, every available pocket is stuffed with disarticulated shells chosen only on the basis of completeness. I sit down on the beach, empty my pockets, and begin to sort the shells into piles separated by species. From my census, the largest number (25) belong to the bittersweet shell (*Glycymeris maculata*), with its distinctive brown-orange swatches. Also well represented in number (20) is the chocolate shell (*Megapitaria squalida*) with its creamy-brown streaks. Third in number (11) is the ark shell (*Anadara grandis*) with its heavy ribs. These are the primary denizens of the sand flats north of the island.

A low rampart of fine sandstone strata that edges the beach separates two dune fields on either side. These orange-brown strata are eroded by the waves during the highest tides (see figure 34) and probably extend seaward beneath the shell beach. In all likelihood, the ground now occupied by the dunes was scooped out during the last relative high stand in sea level some 22,000 years ago. The strata appear to be flat lying at the beachfront when viewed end-on, but they form a low ramp climbing inland where they abut against Miocene andesite. We are flummoxed by the appearance of these strata, because we have encountered nothing like them elsewhere on the island. To start with, they are composed of sand and silt, and there are no discernible fossils to be found. It is only on reaching the far west end of the beach that Dave finds a layer with some large pecten shells, disarticulated, that can be attributed to an early Pliocene species, *Flabellipecten bösei* (Carreño and Smith 2007, 87). Most intriguing, the shells retain a loose calcareous fill composed almost entirely of microfossils. Being careful to keep the filling intact, he extracts some of the pectens. The shells and contents are wrapped securely in tissue paper and stuffed in a ziplock bag together with some notes jotted on a slip of paper. It will not be possible to identify the microfossils until we return Stateside. I record in my notebook that the strata around the layer from which Dave extracted the shells dip 12° to the east. It is an odd inclination for Pliocene strata on this island. I suspect that the angle simply reflects the steeper margin of a small basin. How these strata fit with the prevailing limestone strata we have tracked elsewhere on the island is a mystery left unsolved for the time being.

Distractions on our traverse across the beach have absorbed an hour's time, and the sun is notably declined toward the horizon. We decide to try a shortcut around the northwest tip of the island (see map 9). The route takes us uphill almost 100 feet (30 m) on an easy climb. Once we gain sight of the island's west shore and start our descent, however, we are in trouble. We have wandered into a maze of pitahaya cactus (*Machaerocereus gummosus*). The plants form a dense thicket, spreading profusely from points where stem tips sprawling on the ground take root. We do our best to keep on a downward path angled to the coast, but now and again we must retreat upslope to find an alternative path through the spaghetti web of prickly cacti. I cannot help but think about the web of information on the island's geology that we have gathered on this outing. The simple facts we have collected are raw and undigested and will take more time to sort out. Thankfully, there is no pressing deadline for that. At last we push our way through to the wave-cut platform, agreeably devoid of plant life beneath the sea cliffs.

It is now 4:25 p.m.; the tide is rising but leaves enough space for us to charge south. We cross the next 2.2 miles (3.5 km) along the platform in 50 minutes, a record for geologists who normally pride themselves in taking the necessary time for careful observations. We had walked this stretch earlier in the day, and all is familiar. On reaching limestone that crops out along the shore, we must make a decision to stay with the wave-cut platform or strike inland. Ahead, a jumble of large limestone blocks obstructs the coastal route. With the tide rising, we would be forced to climb over the mess or wade into the water to reach the other side. We choose the inland route, which entails a rapid climb of nearly 200 feet (60 m) to flats that gradually descend toward camp. Thickets of wiry palo adán demand that we take extra caution to avoid long spines, and our progress is slowed. It is startling to see these desert plants thrive on a former seabed of limestone crowded with the shells of fossil pectens. Their plentitude once supported a huge commercial enterprise that employed platoons of helmet-divers to collect scallops from depths commonly between 50 and 60 feet (15.25–18.25 m)[6] on shelves around gulf islands not unlike Monserrat.

The sun has set, and a rosy afterglow bathes the shore as we enter the campsite to be reunited with our waiting team members. There is much animation in the back-and-forth discussion of the day's events. My muscles ache, and it feels as if a great burden is lifted from my shoulders when I remove my backpack. I spy a large pot covered by a lid sitting on our camp stove and inquire about prospects for the evening's dinner. "Check it out for yourself," says Elizabeth with a wry smile. I am too eager to even notice that the stove is not turned on. Lifting the lid, I am assaulted by two Sally Lightfoot crabs (*Graspus graspus*) that attempt to escape their prison with wild leaps from the pot's shallow pool. My reaction is like opening a jack-in-the-box, and the crew erupts in laughter. The students are as excited about telling how they managed to catch the crabs unharmed as Dave and I are about relating our adventure on the limestone plateau. We will strike camp and return to Loreto in the morning.

Various scientific results from the 2006 excursion to Isla Monserrat emerged in due course. In consultation with experts on microfossils,[7] Dave learned that the filling from his pecten shells included numerous species of calcareous nanoplankton. They represent a range of single-celled organisms capable of photosynthesis. Among them are those that delimit the deposit to an age between 4.2 and 3.8 million years (*Rabdosphaera porcera* and *Reticulofenestra minuta*). It means that the sandy Pliocene strata on the island's north end clearly predate the more fossil-rich limestone beds we devoted much of our time to studying. Because the layers with the pectens and associated microfossils dip below the cliffs at the center of the beach, they represent the oldest sedimentary strata on the island.

Jorge and his graduate student Astrid refined the story regarding thick pecten beds in the Pliocene strata from the southwest corner of the island near camp. A relatively rare pecten (subspecies *Patinopecten bakeri diazi*) proved key to delimiting a mid-Pliocene age for that deposit. More striking, masses of fossil pectens shaped into spherical structures likened to armored mud balls were found near the lower end of exposures with strata conforming to a 6° ramp. The working hypothesis on these structures,[8] up to 32 inches (80 cm) in diameter, argues that they formed during a submarine slump in much the same way that a snowball grows by accretion while rolling down a hill. The glue holding the pecten balls together was sticky clay. Marine slumps are normally triggered by a shock wave from a volcanic eruption or an earthquake. Faults on Isla Monserrat were reactivated during the deposition of Pliocene strata (Johnson, Ledesma-Vázquez, and Montiel-Boehringer 2009). Vertical movements that continued well after the last Pliocene strata were deposited on the island clearly forced the significant uplift of those strata that are so evident today. Our final overview was all about mending those ruptures to restore the island to its time of maximum flooding beneath the waves.

On commercial service from La Paz to Los Angeles, the pilot often follows an afternoon flight plan that passes near Isla Monserrat. One must anticipate the approach, because the sighting can be very brief. When the opportunity presents itself, I am grateful for the reunion. The scene from on high (see plate 14) captures the jagged, fault-dissected landscape that so appropriately gives the island its name. Accented in white, the table-like patches of Pliocene limestone that drape parts of the island stand out in dramatic relief. In total, the surviving Pliocene capstone has been reduced by erosion to a mere 9 percent of the island's surface area. The largest remaining sector is the plateau that Dave and I explored, which amounts to 88 acres (356,000 m²) in size. On those occasions when the island is visible, I am instantly reminded of the physical trials we endured to get there and the insights retrieved through a determined effort. I reimagine the protector of Monserrat, the last Amazon of Juanaloa, and I celebrate her shielding spirit.

9

Riding Out Ancient Storms on Isla Cerralvo

Every traveler has a right to invent his own geographies.
If he didn't, he would be no more than a traveller's
apprentice, still bound by what his teacher taught him.
—*José Saramago,* Journey to Portugal

THE POWER OF A PHOTOGRAPH to entice a traveler into visiting a new landscape should not be dismissed lightly. Images are seductive, even if things are not as obvious as they might otherwise seem on the surface. We sometimes go places for the first time thinking we already understand what awaits us, even if the journey's purpose is to tease out the finer details. The grainy photograph that brought me to Isla Cerralvo appeared in a report issued by the California Academy of Sciences some 42 years before I first arrived on the scene where the photo was taken, roughly 26 miles (42 km) east of La Paz. The research article was written by a paleontologist (Hertlein 1966), who described Pliocene fossils from localities in the southern part of the peninsula. It was not the invertebrate fossils that drew me, but rather the photo of a sea cliff on the west side of Isla Cerralvo. According to the photo's caption, a "white stratum" in the middle of the picture was composed of "calcareous algae of Pliocene age." In the photo or otherwise provided in the caption, there is no hint whereby the thickness of the stratum might be appraised. However, the limestone layer was likely to be formed by extensive rhodolith debris of the kind encountered by us on Coronados and Monserrat islands (see chapters 7 and 8). A trip to the lower Gulf of California to explore the extent of more-southerly deposits influenced by rhodolith input seemed perfectly in order.

On closer reading, the academy paper revealed that the author had not visited Cerralvo himself but acquired the fossils and the photo from third parties. Most intriguing, the paper's acknowledgments also relate that the original collecting trip was made possible through the generosity of Mr. Crosby, who furnished his yacht *True Love* for transportation.[1] Whether the research party (minus Hertlein) traveled all the way from Los Angeles to Isla Cerralvo aboard the singer's yacht is not clear, but one can only marvel at the prospect of mixing personalities from the scientific and elite entertainment worlds. In my wildest musings, I never dared to think of a benefactor with the deep resources to mount a trip by yacht through the Sea of Cortez. The only way I was sure to get a student team onto Isla Cerralvo was by open panga.

As is so often the case, Jorge had excellent contacts in the region and found a ready-made solution for our supply and transportation requirements. American expat Tim Hatler had settled in the village of La Ventana, located on the peninsular mainland across from Cerralvo, where he operated a small resort catering to divers, fishermen, and kite-surfers. The connection was that Tim married a local girl, Jimena, who happened to have taken courses in marine science with Jorge as an undergraduate student at the university in Ensenada. After Hurricane John struck the Baja California peninsula in September 2006, the couple asked the professor to pay a visit and make an assessment regarding erosion in an arroyo running through the property. If unchecked, additional collapse of the embankments threatened to undermine valuable infrastructure. Jorge had not gone out to Cerralvo (5.5 mi. or 8.75 km offshore) during his initial visit, but he came away with the impression that Tim could assist us if we desired to pursue studies anywhere in the immediate region.

The notion of hurricane-strength storms as a source of landscape modification was not fully on my radar during our earlier studies up and down the Gulf of California. Although wind typically played a role in weather conditions during our January visits, the skies were usually clear and sunny. The notoriety of major storms on a frequency every few years apart was kept out of mind. Jorge's mission in visiting La Ventana in the first place should have been my biggest clue of things to come. Once in contact with Tim, however, I felt that the problem of linking the 1964 photograph to a concrete locality on the island was paramount. According to the photo caption, the locality was at a place called El Mostrador. Tim assured me that nothing like the description of extensive white cliffs was at El Mostrador. He thought what we were looking for would be found at a place farther south called Paredones Blancos (or "white bluffs"). It seemed that we might be led on something of a fishing expedition, but Tim agreed to meet Dave, two students, and me at the central bus station in La Paz at midday on Tuesday, January 8, 2008, and bring us out to La Ventana.

I will never forget my first impressions of the island, the sixth largest in the Gulf of California, covering an area of 4 square miles (10.46 km^2). The paved road to La Ventana from La Paz cuts inland behind high coastal ridges, only to regain sight of the gulf on high ground some 5 miles (8 km) west of the village (see map 1, locality 9). Tim pulled his big Carryall Suburban to the side of the road to let us stretch our legs and enjoy the view. The first thing to hit me was the island's size and layout. From where we stood, it was possible to see only the southern half of the island. Cerralvo has an elongated shape, stretching over a central ridge nearly 18 miles long (29 km). Like a talking atlas, Tim informed us that the highest point on the ridge stood 2,530 feet (771 m) above sea level (map 10). Mysterious Monserrat is an island that could be penetrated on foot with hard work, but how would we manage to conquer Cerralvo? Under the early afternoon sun, whitecaps on the Cerralvo Channel were the next features I noticed. The north wind shot through the passage, kicking up swells crested by white spray. Indeed, a good part of Tim's business catered to the kite-boarders attracted to this very area on account of the winter winds. He proudly pointed out numerous colorful kites closer to the mainland, each attached by long lines to figures standing on what looked like surfboards. Now and again, those figures would leap into the air off one of the swells. Tim began to tell us about the annual kite-board regatta that would be held in a few weeks' time, a race from La Ventana to the beach near the light tower at Punta Sudoeste and back again. The event apparently drew top competitors from all over North America and abroad.

Between La Ventana and the island, the channel water was dark blue in color brushed by lacy whitecaps. Exactly how deep was the water? I wondered out loud. Happy to oblige with facts, Tim informed us that the deepest part of the channel went to about

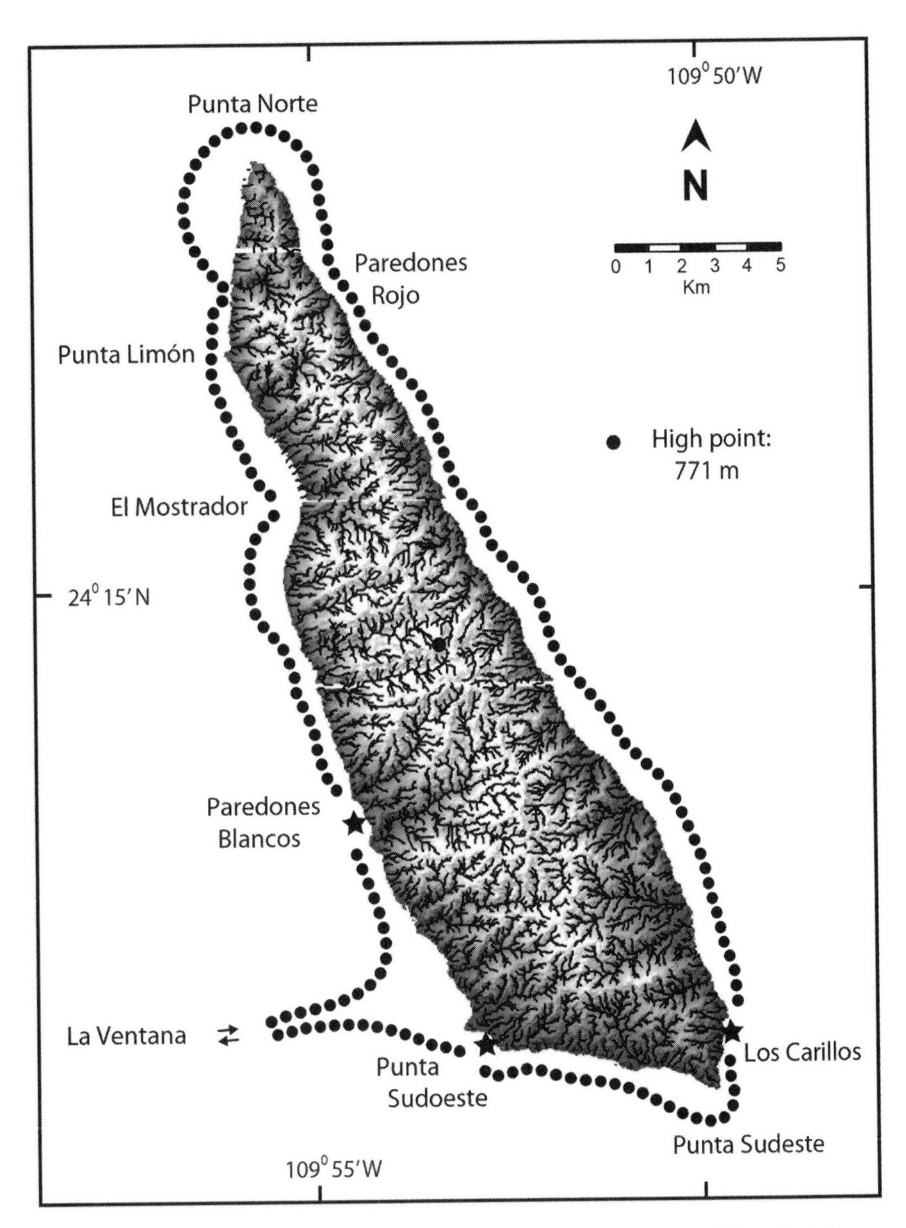

Map 10. Topography of Isla Cerralvo. *Note:* Heavy dotted line (with directional arrows) shows boat route; black stars mark stops ashore. Map by author and David H. Backus.

1,700 feet (518 m) but that the depth dropped to 2,400 feet (732 m) at the north end.[2] A shallower shelf extended from the south end of the island to Punta Arena some six and a quarter miles (10 km) east of La Ventana, rarely dropping below 600 feet (183 m). Thus, the trough separating Cerralvo from the peninsular mainland is roughly two-thirds as deep as the maximum height to which the island rises. The contrast made sense in comparison with other gulf islands such as Angel de la Guarda, Carmen, and Monserrat that stand as coherent horst blocks separated from the Baja California peninsula by sunken grabens. I was beginning to take the measure of the place.

By the next day, we were encamped adjacent to Paredones Blancos with a knowledgeable boatman and provisions for several days. It was the first of three trips to the island, with others to follow over the next two years. Cerralvo proved to record more than one good story regarding fossil rhodoliths, but a serious contradiction to the facts eluded me at the time. The center of Isla Cerralvo is fixed on a line of latitude less than a full degree north of the Tropic of Cancer, a position that qualifies the place as a near tenant of the tropics. Limestone ought to be everywhere. As we were soon to learn, the island held additional attractions for geologists, particularly in terms of landscape evolution. In the end, our group published several reports regarding the paleontology, geology, and landscapes of Isla Cerralvo (Backus et al. 2012; Emhoff et al. 2012; Johnson et al. 2012; Tierney and Johnson 2012).

In practice, any camping on Isla Cerralvo or extended visits require permits and the consent of individual landowners. The visit outlined in the following itinerary (see map 10) is designed as a full-day trip from La Ventana that makes a circumnavigation of the island by boat through a distance of 40 miles (65 km), not including transit of the Cerralvo Channel. Many important features of the island may be comfortably viewed from a fisherman's panga, abetted by three or four landings to more closely examine coastal sediments and rocks.

FEATURE EVENT

Tracking Old Hurricanes on Isla Cerralvo

It's an early breakfast, finished hastily in time to herd the crew down to the beach where our boatman waits. Food and refreshments for the day have been prepacked in a cooler and already loaded aboard. We want to shove off during the morning calm before the usual northerly wind revives. Our guide is an experienced local man who knows the place well after years of hunting wild goats on the island and fishing all the good spots around. Our course is set toward the light tower at Punta Sudoeste. The sea is flat, and the boat fairly skims across the surface. It will be a dry passage, although the precaution is taken to secure boots and socks in a waterproof bag. On our approach toward a vast sand beach, we veer north, drawing closer to the granite coastline. Our first landing will be at Paredones Blancos (see map 10). For the time being, everything to the starboard side is dominated by igneous rock punctuated by small outlets where arroyos meet the sea to build close-hemmed sandy beaches. Mostly, the granite dates from Cretaceous magma chambers unroofed probably in the late Miocene.

After we cross a wide arroyo mouth, the first signs of white rocks appear in the sea cliffs. It looks as if there is not one great stratum, but several recurrent layers of limestone. This is not what we anticipated from the 1964 photo. I ask the boatman to slow to a crawl and move the boat closer in to shore. The white layers are contorted and shot through by faults. The repetition of layers is a visual illusion caused by the offset of normal faults. Slowly, we drift alongside a more coherent part of the fault block. What we now find makes a credible match with the 1964 photo. We pull farther north toward a sandy stretch of beach where we can disembark. The view back to the southeast puts the "white stratum" within its larger fault block into perspective (see plate 15). Draped by morning shade, a narrow arroyo intersects the coast. A few cardóns grow at the arroyo's mouth. They are easily 20 feet (6 m) in height. The stature of these cacti provides a useful measuring stick against which to judge the enormous size of boulders calved from the outcrop. Indeed, this is the place. The photo caption in the Hertlein (1966) paper was correct in describing a single, great white stratum.

The shore is littered with a bewildering array of cobbles and small boulders. The first order of business is to find a dry spot and put on my field boots. As I squirm my cold feet into the boots and lace up, I cannot help but notice that the cobbles around me include pieces of granodiorite (pink with abundant plagioclase), hornblende diorite, and basalt (dark and heavy), as well as foliated gneiss (a metamorphic rock with wavy bands of mineralization). All reached the shore via the small arroyo behind us, and they tell us about the island's interior composition. We will be here awhile before continuing around the island. I need to confirm that the white stratum is dominated by rhodolith debris.

The scale of the outcrop presents a challenge. It will not be easy to get at the sheer cliff face where the white limestone is exposed in place. Initial impressions are based on the contents of the huge pieces calved like icebergs from the cliff face. There is a subtle layering within the great seam, and one can easily see that some hunks are turned askew, while others essentially slid straight down the cliff to lodge in an upright position, retaining a semblance of stratigraphic order (oldest part on the bottom, youngest on top). From top to bottom, much of the stratum is dominated by finely comminuted rhodolith debris. Broken bits of delicate branches and the bud-like tips of the coralline red algae are densely packed together (figure 35). At least initially, there is no sign of whole rhodoliths preserved intact.

Figure 35. Coarse debris derived from crushed rhodoliths in the middle of great white seam at Paredones Blancos (scale 4 cm). Photo by author.

While other crew members turn their attention to the massive conglomeratic stratum below the white seam, I am obsessed with getting a more accurate estimate of the seam's true thickness. With a cardón rib in hand as a suitable staff, I struggle up the scree at the far end of the outcrop to reach the contact between the base of the limestone and the underlying conglomerate. I dare not climb any higher. Standing at the contact, I raise my arms as high as possible and extend the staff straight above my head. Those watching below agree that the tip of the staff meets a level equal to the top of the seam. Before rejoining the crew, I take a quick look at the bottom part of the seam and recognize that it includes a few whole rhodoliths mixed together with sparse fossil coral fragments, broken pieces of sea urchin tests, and pecten valves. Inclusion of these fossils signifies that something of a transition exists to the pure rhodolith detritus. Once down from my perch, I lie flat on the beach and strike the same pose with hands over my head and staff extended beyond. Geology student Kristen Emhoff stretches a measuring tape from my feet to the tip of the staff and marks off the distance. It is close to 33 feet (10 m), the thickness of the great white stratum.

Discussion now focuses on the nature of the conglomeratic beds below the white stratum. Although the base of this thick unit is hidden by scree including the fallen blocks of rhodolith limestone (see plate 15), access to the upper part is good. We soon understand that the range of rock types represented by clasts within the conglomerate is the same as that found at the mouth of the modern arroyo. Variation in the size of cobbles and boulders also is comparable to that of those found on the shore. As with the modern setting, there is a liberal amount of silica sand included in the matrix to the conglomerate. We search, but no marine fossils are preserved in the matrix. We conclude that the conglomeratic unit represents a terrestrial environment basically identical to that forming now at the mouth of the arroyo. In stratigraphic position undisturbed by faults, there occurs a second conglomeratic unit above the limestone. On the whole, the succession of strata captured within the fault block signals the marine flooding of a stream valley during which the rhodolith material was introduced. Subsequent to that, a relative fall in sea level appears to have reinstated the former environment with the progradation of coarse material eroded through the drainage.

Taking things a step farther, the top of the fault block is crowned by another sort of limestone. As we make our way back toward the boat, we find boulder-size blocks of well-cemented limestone with fossil corals of a branching morphology that have tumbled from above or were carried downstream from a more inland locality. The branches are often intact and about the same size as those from the Pleistocene reef on Isla Coronados (see figure 28). It is a crude stratigraphy but one that gives a basic outline to the order of events in the struggle between land and sea. Enough is here for several days of work, but we must press on to complete our tour of the island's perimeter.

Once we are relaunched, we must draw out into the channel to bypass a great bulge of coarsely mixed sand, cobbles, and boulders left exposed by the low tide. The detour brings us nearly a half mile (800 m) to the north and across another wide arroyo mouth. Penetrating well inland, the arroyo exhibits a surprising gradient, perhaps as high as 8 percent, rising from the shore. The bulge of sand and cobbles does not extend directly out to sea from the arroyo mouth but is skewed to the south. Past the beach, we are back in granite territory coasting close beside the shore. I remain suspicious regarding the original site description for the cliff of rhodolith limestone given by Hertlein (1966), and I am eager to reach El Mostrador (see map 10). We pass part of the coast where the rocks shift from pink granite to darker metamorphic rocks, possibly the same gneissic rocks that are found as cobbles in the massive conglomerate at Paredones Blancos.

Soon, we round the coast below El Mostrador. A ranch with a few cattle was located here some time ago, but no trace of buildings or a corral survives. I signal with downturned palms of my hands for the boatman to slow our progress. With binoculars, I scan the slopes above the shore for any trace of a color change that might indicate limestone. Nothing but granite is in sight. Staying close to shore, we resume our speed north toward the bump of land projecting from the shore at Punta Limón. No limestone reveals itself along our course.

Rocky islets poke above the waves off Punta Limón. We pass them inboard and follow the island's profile along a more north-south section of the coast. The wind is picking up to push sea swells. Near its zenith in the January sky, the sun informs us that the busy morning has all but vanished. It is time to make landfall for a lunch break. There are more arroyos that cut drainages through the landscape to bring copious supplies of sand and cobbles to the coast. The one ahead intersects the rocky shore in such a way as to offer a corner of shade. Dave signals for the boatman to turn shoreward, and we head for the middle of the arroyo mouth (see map 10). The tide has only begun to rise, and the boat slides onto the beach below a vertical wall of coarse sand, four feet (1.22 m) high. It is an odd sort of barrier not at all typical of a beach profile, looking almost as if it was neatly carved by a great scalpel. Some effort is required to climb over the barrier and move the cooler with our lunch supplies onto the upper beach. Underfoot, the sand is still damp. We are reduced to scooping out a ramp using the flat edge of our geology hammers to cross the barrier. Here, too, the slope of the arroyo floor rising inland seems oddly steep where it merges with the upper beach.

Viewed from our position comfortably seated in the shade, various details of our lunch spot fall into place as part of a larger pattern. The peculiar warping of the sandy apron washed beyond the beach onto the tidal flats reconfirms evidence observed earlier in the day. The apron is not perfectly symmetrical but lopsided toward the south. Although whitecaps have yet to appear in the channel, the passage of ocean swells to the south is a vivid reminder of the winter wind. The vertical sand barrier we labored to surmount was brushed by the last high tide. Coarse sand undercut by the rising water had slumped onto the shore face where it was swept away by a long-shore current pushed by winds from the north. Sediments so removed were contributed to the seaward lobe of sand drifting to the south adjacent to the rocky shore. Here, then, is reaffirmation of a model for physical processes in operation around the island on a daily basis during the winter months.

It occurs to us that the huge sand bulge near the south end of Isla Cerralvo above Punta Sudoeste, where we turned earlier in the day (see map 10), is the product of sand collecting from the outwash of a much larger arroyo system. Would the same pattern hold up on the east side of the island? Although disappointed by the lack of limestone north of Paredones Blancos, we now have a new story line to consider. *River delta* is the most appropriate geographical term to broadly fit the details of our experiences on the west shore. As an allusion to the triangular shape of the Greek letter *delta*, the word harkens back to usages by the early Greeks to describe river mouths. The most famous of such appellations was recorded by the historian Herodotus in reference to the Nile River delta sometime around 440 BCE: "The Delta is Egypt," he wrote.[3] But there is nothing green and fertile about the landscape surrounding our lunch spot. Where on earth does the water come from that brings all the coarse sediments down to the edge of the sea on this island in the lower Gulf of California?

With this open-ended question in mind, the crew prepares to reboard the panga and continue our reconnaissance. To the west, some 25 miles (40 km) distant, is the south tip of Isla Espiritu Santo. The sun shines brilliantly. The cloudless sky is pale blue above dark-blue water, and our open vessel labors northward against the sea swell. Rounding Punta Norte is not a simple matter. Small islets and sea stacks are hazardous obstacles with threatening waves that surge against the rocks in foamy fury. Our boatman follows a wide arc around the north end (see map 10), taking us well outside the tumult of the waves. In making the turn, however, we run eastward for several minutes athwart the swells. The panga rises uneasily on each swell and sinks sideways into the trough as the bulge of water passes beneath. The day is resplendent in sunshine, but my stomach sinks faster than the rest of my body with each pitch of the boat. For the briefest moment, I wonder what such a land-loving geologist as myself is doing in this small boat. Soon enough, our able boatman brings us around on a south course. Now the wind is behind us and the panga picks up a rhythmic surge with each swell. It will be a faster ride along the outer coast to the south end.

On his own accord, the boatman soon eases the throttle and points to the high upper cliffs some 300 feet (90 m) above the water-line. We have reached the Paredones Rojo, or "red bluffs" (see map 10), a distinctive local landmark. These match the pink rock that is used as a facing stone on many buildings around La Paz. A large quarry on the edge of town off the paved road leading to La Ventana supplies the material. The capstone is rhyolite, the volcanic fine-grained equivalent of granite. Rhyolite must rise through the neck of a volcano to reach the surface and spread across the landscape as a flow. Unless tilted sometime afterward, rhyolitic flows remain relatively flat. Although the widespread granite in the cape region of the Baja California peninsula is essentially the same Cretaceous granite as found in the Cataviña Boulder Field far to the north (see chapter 1), the regional presence is characterized as belonging to a separate terrain given the formal name Pericú (Sedlock et al. 1993). The stocks, or volcanic necks that cut through the cape granite, clearly came afterward and are thought to be vaguely Tertiary in age.

Resuming speed, we remain vigilant for any enclaves with limestone or any arroyos that intersect the shore to deposit deltas truncated by the vigorous long-shore current. Theoretically, a relative rise in sea level might flood the valley openings with marine sediment enriched by rhodolith debris. That is the working hypothesis formulated at Paredones Blancos, and there is every reason to think the same process should have functioned on the east side of the island. The valley mouth occupied by marine sediments of the white bluffs was unusually broad for the island, essentially the outcrop length of Paredones Blancos (a distance of almost a half mile, or 0.75 km). A promising locality appears ahead. Dave signals the boatman to slow, but the current urging us along the shore makes it difficult to come to a dead stop. The boatman swings the craft around, pointing up the current at an oblique angle to the coast. With the engine still engaged to provide stable forward momentum, the panga gently edges toward the shore. Around us in the relative stillness, the first whitecaps of the afternoon barely ruffle the water's surface.

A pair of narrow arroyos entrenched on the flank of the steep slope attract our attention. Here, the bedrock is degraded and the ground is held in place by a forest of medium-tall cardóns interspersed by acacia trees. Compared to arroyos observed on the west side of the island, these are shorter with a steeper gradient. All the same, they open onto the shore with small but pronounced deltas that splay out symmetrically above the beach (figure 36). As seen during lunch, these fan deltas also are decapitated by a combination of tides and strong long-shore currents that mine the coarse sediments and move them, conveyor-like, southward over the subtidal shelf. The neighboring fans are close enough to merge together. The last high tide left a vertical wall of sediment, more elevated in the center but tapered lower to the lateral extremities of the adjacent fans. Using the ubiquitous cardóns as measuring sticks, we estimate that the cuts reach a maximum height of five feet (1.5 m). This does not mean that the tide level rose so much, only that the moving water undermined the toes of the fans. Since Herodotus, physical geographers have erected a box-work classification of deltas to highlight their

Figure 36. Pair of small arroyos and connected fan deltas on the east coast of Isla Cerralvo. The cardón cacti on the slope are up to 10 feet (3 m) tall. Photo by author.

variations. It strikes me that the small deltas in front of us represent shoal-water fans with a moderate gradient.[4] Basically, the imposing island of Cerralvo is attacked from all sides, slowly reduced in stature by one eroded grain of sand, one pebble, and one cobble at a time.

Nowhere, yet, have we encountered rocks composed of limestone along this coast. The shelf on the island's east flank is a narrow one that rapidly drops into a neighboring trough called the Cerralvo Basin (Dauphin and Ness 1991). All we see to the east is dark-blue water. Four miles offshore (6.5 km), the basin attains a depth of 5,000 feet (1,524 m). If the center axis of the island jumped eastward by a relatively short distance, the entire place would disappear into the basin with enough room left over for another island of equal size. The Mexican mainland is out of sight, eight hours away by car ferry from La Paz. Thoughts on how marine life might be prohibited along these shores begin to percolate through my mind. The narrow shelves on both sides of Isla Cerralvo do not have much space for the kinds of bottom-dwelling organisms that commonly use calcium carbonate in shells, tests, and algae. The marine life that does so has a preference for clear water unclouded by clastic sediments.

The next spot where the panga can be pulled safely ashore is at Los Carillos, a short distance above the island's southeast point (see map 10). It is nearly 2 p.m., and the opportunity to stretch and to explore the shore is a welcome relief. A generous sand beach rises sheltered behind a rib of strata fully submerged only during the highest tides. In season, the place is known as a nesting ground for green sea turtles (*Chelonia mydas*). Our boatman has been here many times. Under the rising tide, waves dash at an oblique angle against the outer rocks while the beach on the other side remains unaffected. The boatman brings us to the far end of the rocks and makes a sharp turn into a small channel. No time is wasted jumping clear to pull the panga securely onto the sand.

The rocks shielding the beach consist of conglomeratic layers, some with small pebbles all the same size but others with a coarser mixture of cobbles and small boulders held in a strongly cemented matrix of sand. There seem to be no fossils, but Dave continues to explore layers dipping steeply outward toward the surf. Looking south along the coast, I spy strata that are more gently dipping above the tide line. To get there, it will be necessary to pass over a narrow ledge around a cliff that shuts off our end of the beach. The cliff is an extension of the conglomerate layers enclosing the beach, also entirely devoid of fossils. If the tide rises too far, returning by the same path will not be possible. I motion to our boatman that I am prepared to cross over. He glances at his watch and raises an index finger with the admonition "Una hora." It will be a hasty tour, but the prospect of limestone is a powerful draw.

Joined by Kristen, we quickly ascertain that a sheet of coarse-grained sandstone covers calcareous sandstone with scattered pectens but also a few whole rhodoliths. Less than 30 inches (76 cm) thick, these layers rest on a thick conglomerate formed by granite cobbles with an admixture of coarse sand and rare basalt pebbles. The conglomerate also includes fossil gastropods (*Nerita scabricosta*) and a smattering of worn pecten shells (figure 37). An uneven granite floor supports the conglomerate and lies exposed where the waves have stripped away the covering strata. The capstone dips seaward about 7°, a slope that is typical for a ramp deposit that accumulated in place because of onlap of marine sediments (see plate 16).

Wandering off the wedge of conglomerate seaward where the tide has yet to return, I am startled by what I find and give out an audible gasp. Kristen joins me, and we crouch together on the rocks for a

Figure 37. Pliocene conglomerate at Los Carillos with fossil gastropods (*Nerita scabricosta*) and worn pecten shells. Photo by author.

closer view. Where we sit at the base of the conglomerate, granite boulders are welded together directly atop the unconformity surface. Thousands of rhodoliths somewhat smaller than a golf ball in size fill the spaces among neighboring boulders (figure 38) and drape across other boulders to a thickness of three or four rhodoliths. Individual spheres are firmly cemented together, but they are also badly worn by the daily scour of the tides. Sufficient detail remains, however, to show that many have a lumpy morphology with compact, sturdy branches ending in thickened tips. A hurried survey demonstrates that few other fossils are mixed into the deposit in an area covering roughly 250 square feet (23 m^2). Some fossil oyster shells are attached to larger boulders, but more generally the broken shells of pectens are rare among the mass of rhodoliths. Encrusting oysters and the many neritid gastropods that characterize the overlying conglomerate speak convincingly to an association of organisms that lived on an active rocky shore. In contrast to the fossil rhodoliths discovered

Figure 38. Fossil rhodoliths spread among granite boulders at Los Carillos. Pocketknife (3.5 in. or 9 cm long) for scale together with carapace of a Sally Lightfoot crab. Photo by author.

on the limestone plateau of Isla Monserrat (chapter 8), these seem out of place.

As with fossils preserved in rocks of any kind, the distinction must be drawn between organisms buried in the setting where they lived, in contrast to those removed from their original environment and buried elsewhere. Regarding the life and vagrancies of rhodoliths, Charles Lyell would have us check to see whether any occur today on rocky shores in an intertidal setting. While there is some evidence that rhodoliths may live on tidal flats (Perry 2005), that is far from normal. In any case, the fossil rhodolith bank left undisturbed on Isla Monserrat occurs one and a quarter miles (2 km) from the nearest outcrop representing a rocky shore of the same geologic age.

According to Steller and colleagues (2009), dense aggregates of rhodoliths living in the Gulf of California occur today at depths usually no shallower than about six and a half feet (2 m). The fossil rhodoliths here at Los Carillos must represent a death deposit created when a storm with powerful waves swept them onto the granite shore. Hurricane Marty in 2003 accomplished the same thing, stranding a mass of living rhodoliths on the coast near Punta el Bajo north of Loreto (chapter 7). If not pulled back off the shore by a powerful undertow, the coralline red algae would bleach under the sun and die. How might such a rare occurrence from a single storm be left intact and become preserved in the rock record? The details are here, somewhere. A loud whistle carries across the water from the beach. The boatman waves and beckons us to return.

To reach the boat, we are forced to dash across an open ledge during brief intervals between waves. In the rush to launch the boat, our eagerness to share information with Dave and to learn what he has found is chaotic. Lacking any fossils, the stratified rocks fronting the beach are peculiar. The absence of fossils is perplexing. Dip angles vary from spot to spot, and all are steeper than the unified carbonate ramp that advanced over the adjacent rocky shore, probably during the Pliocene. But Dave gives me a grin that lets me know he has some inkling about the noncarbonate rocks fronting the beach.

Under the whine of the outboard engine, carrying on a conversation without shouting is all but impossible. As we turn out from the beach and continue south, my eyes are glued to the shore. Although the tide is higher than before, I spy a basalt dike that cuts through the granite as a vertical wall. It must be 6.5 to 10 feet (2–3 m) wide, and it rises above the surface of the granite. The dike is partially covered by limestone on dry land. On second look, there is more than one dike, all running out to sea roughly parallel to each another. Minutes later, we round the island's southeast point where sea lions (*Zalophus californianus*) sun themselves.

Our course is now due west. Almost immediately, the sea swells are gone and the water is becalmed. We have reached the short, leeward end of the island out of the winter wind. Along this stretch, the widest part of the island, all arroyos drain southward over a more subdued landscape. Dave gestures inland toward a cluster of brown hills surrounded by the pink overtone that is typical for granite. Out of the wind, exchanging words is a little easier. Dave growls two words: "Thermal springs." The coloration is reminiscent of the fossil hot springs that line margins of the great Pliocene bay at San Francisquito (see chapter 4). Those former springs undoubtedly played some role in weakening basement rocks that led to the erosion of the embayment. It is logical to suppose that similar limestone might be found here like fingers creeping up some of the eroded drainages rotted by mineral-rich water. All eyes train intently on the coast, finding no trace of limestone.

When we round the island's southeast corner, the story quickly changes. Slabs of limestone rest slumped below the steel light tower some 20 feet (6 m) above the water. It will be the last stop of the afternoon before we recross the channel to La Ventana. Our boatman brings the panga through a wide turn, steering for a broad swath of sandy beach. Although back on the west side of the island, we are sheltered from the wind by high sand dunes to the north. The dunes bulge like a flexed muscle against the flow of the long-shore current streaming down from the north. South-directed swells near the shore are refracted eastward around the bulge, giving the boat an added push onto the beach.

Limestone layers that project from the south end of the beach are inaccessible, plunging into deep water. To reach a convenient ledge where we might examine the contents more closely, we must take a detour toward the light tower. At the top of the rise, the ground is littered with Pleistocene corals and shells weathered loose on the surface. Our goal is to reach the solid strata below. It is an easy descent, first to a wide ledge about eight feet (2.5 m) above the water and then to a lower, narrower ledge near the water. There is room enough for all, although the sensation is a bit like standing on scaffolding. Now and again, gentle waves lap at the far end of the ledge, but the fresco in front of us is so astounding that no one pays any heed.

In a scene reminiscent of the coral reef from the fossil lagoon on Isla Coronados (see chapter 7), we are treated to a line of tall corals perfectly preserved in growth position. Better yet, we stand before a cliff face that exposes a succession of fringing reefs growing one above the other on an open coast. Unlike the wide-branching colonies found on Coronados, these are cone-like in profile and densely crowded. Most are a morphotype of the common *Porites panamensis*; however, some rare branching colonies of *Pocillopora capitata* also occur. Like their relatives on Coronados, almost every colony is attached to a cobble or small boulder. The cobbles on Coronados are all derived from andesite, while those we see here are diverse in origin. Many are granite, others are gneiss, and a few are basalt. More telling, each of the clasts exhibits an enclosing rind of coralline red algae. It would be wrong to say that the coated cobbles are rhodoliths, but the distinction is a fine one based on the ratio between the diameter of the central rock core and the thickness of the enclosing rind. Granite pebbles outside the great Pliocene bay at San Francisquito seeded rhodoliths with rinds up to five-eights of an inch (1.5 cm) thick (see figure 12b). Here, by comparison, the rock cores are much larger, and the algal rinds are far thinner. In both cases, however, the dynamics were essentially the same. Pieces of rock, small or large, were tossed and turned by nearly constant waves while coralline red algae coated them on all sides.

Before coral larvae had the opportunity to settle and grow into large, upright colonies on algae-encrusted clasts, those cobbles and boulders had already ceased to tumble. Another notable difference from the lagoon reef on Coronados concerns biodiversity. Inspection of the many nooks and crannies among the corals in the cliff face reveals a plethora of mollusks that formerly lived within the reef. On and within the corals there is evidence of parasitism. Most obvious are elongated, finger-size cavities left by pholad bivalves capable of boring into corals, while smaller, star-shaped calluses left by boring barnacles are more delicate. The traces made by bivalves belong to a species of *Lithophaga*, while the barnacles represent a species of *Ceratoconcha*. Coming from very different phyla, these invertebrates converged on a similar lifestyle providing maximum protection hidden inside a coral host while living on plankton filtered from seawater imported through small openings to the outside. Better known from the North Atlantic and the Caribbean (Santos et al. 2012), the odd barnacles almost certainly gained a toehold in the Gulf of California prior to about 3.5 million years ago when the Isthmus of Panama emerged to block free passage of marine organisms from opposite sides of the Americas.[5] Surprises of this kind are what make the gulf such a treasure-house of mixed biogeography.

Above, the wider rock ledge exposes a surface more than 300 square feet (325 m^2) in area at a level intersecting one of the fossil reefs. We climb onto the ledge and disperse over the surface searching for hidden fossils, much as if snorkeling on a modern reef. The fossil biota is quintessentially Pleistocene. In the same way a marine biologist might take a census of reef biota by recording data on an underwater slate using a grid, the ledge exposure invites a comparable review using a grid system and an ordinary clipboard. If we have the opportunity to survey a fossil reef and understand the spatial and numerical interrelationships of its inhabitants, then we are also challenged to consider the factors that cause the demise of that ecosystem. On first inspection, it appears that a layer of conglomerate buried each succeeding reef in the profile. Each layer includes the same mixture of clasts derived from granite, basalt, and gneiss. The combination could hardly be produced by normal shore erosion, especially on Cerralvo, where so much of the coastline is dominated by granite. There are too many questions to be answered from so brief a visit.

Already the sun is low against the granite ridge behind La Ventana. There are fewer whitecaps on the Cerralvo Passage, but the swells remain in force. The return trip promises to be a wet one.

During our first excursion to Isla Cerralvo, enough data were collected at Paredones Blancos for geology student Kristen Emhoff to complete an independent study project. Lab work subsequently showed that the purest rhodolith limestone uncontaminated by terrestrial sediments occurs in the middle of the thick seam (Emhoff et al. 2012). A new crew arrived on the island in January 2009, including Jorge, Dave, myself, and my spouse, Gudveig, assisted by geology students Daniel Perez and Peter Tierney. While Dave and Jorge returned to Paredones Blancos to investigate the area's complicated tectonics, the rest of us divided time between Pliocene outcrops at Los Carillos and the younger Pleistocene outcrops at Punta Sudoeste.

Census work at Los Carillos was employed to distinguish between the stranded rhodoliths as a death deposit in contrast to the normal association of fossils in the rest of the conglomerate. A sedimentological analysis devised by Dan also tested how basalt dikes rising above the surface of the granite tidal zone acted as natural groins to keep eroded sediments and storm-delivered rhodoliths in place (Johnson, Perez, and Baarli 2012). Shifting to the Pleistocene reef limestone at Punta Sudoeste, Peter collected data for an honors thesis. A stratigraphic column was compiled detailing repetitive cycles of Pleistocene reef building, each abruptly ended by a deluge of cobbles and small boulders. A comprehensive census of the fossil reef exposed on the big ledge also was made to document the climax stage in ecological succession attained during reef revitalization. Lab analysis of a coral collected from the site gave a radiometric age of 122,143 ± 175 years before present, which confirmed a Late Pleistocene age in close agreement with a date calculated for the lagoon reef on Isla Coronados (see chapter 7). More significant, Peter formulated a hypothesis for how the reef cycles at Punta Sudoeste were impacted by hurricanes that episodically brought huge volumes of rock debris from the interior of the island to a Pleistocene fan delta adjacent to the reef sequence (Tierney and Johnson 2012).

Dave led another boat tour around the island to collect hard data on the modern arroyos and related fan deltas. Information on the extent of living rhodolith banks around the island revealed only a single locality off Punta Limón where they thrive today on a drowned islet. Upon checking the history of recent hurricanes, we found that Isla Cerralvo had taken direct hits on several occasions (figure 39). The most recent had been Hurricane John in 2006, the storm that nearly ruined Tim's enterprise at La Ventana. Based on satellite imagery, a detailed study of the island's drainages and fan deltas became the crowning piece of our investigations on Isla Cerralvo (Backus et al. 2012).

To a degree of severity unknown elsewhere in the gulf, the landscape of Isla Cerralvo is a place where intermittent erosion by hurricane floodwater gradually turns the island inside out through massive transport of sediments to the surrounding shores. Circulation of geothermal waters from the deep crust also plays a part in making the rocks more susceptible to erosion. Between major storms, the normal seasonal work of long-shore currents driven by winter winds and tides redistributes the sediments that are temporarily brought to the island's many fan deltas. Biological productivity of the chief limestone-making biota is choked by suspension of excessive clastic sediment on the narrow shelves around the island. The few Pliocene and Pleistocene occurrences of limestone on the island are limited compared to those of other landscapes explored throughout the gulf. It would be tempting to witness a great storm from the safety of a bunker high above the canyon walls of Isla Cerralvo, but the aftereffects recorded in the rock record are clear enough by inference. It is a case study where the traveler was well schooled by his apprentices.

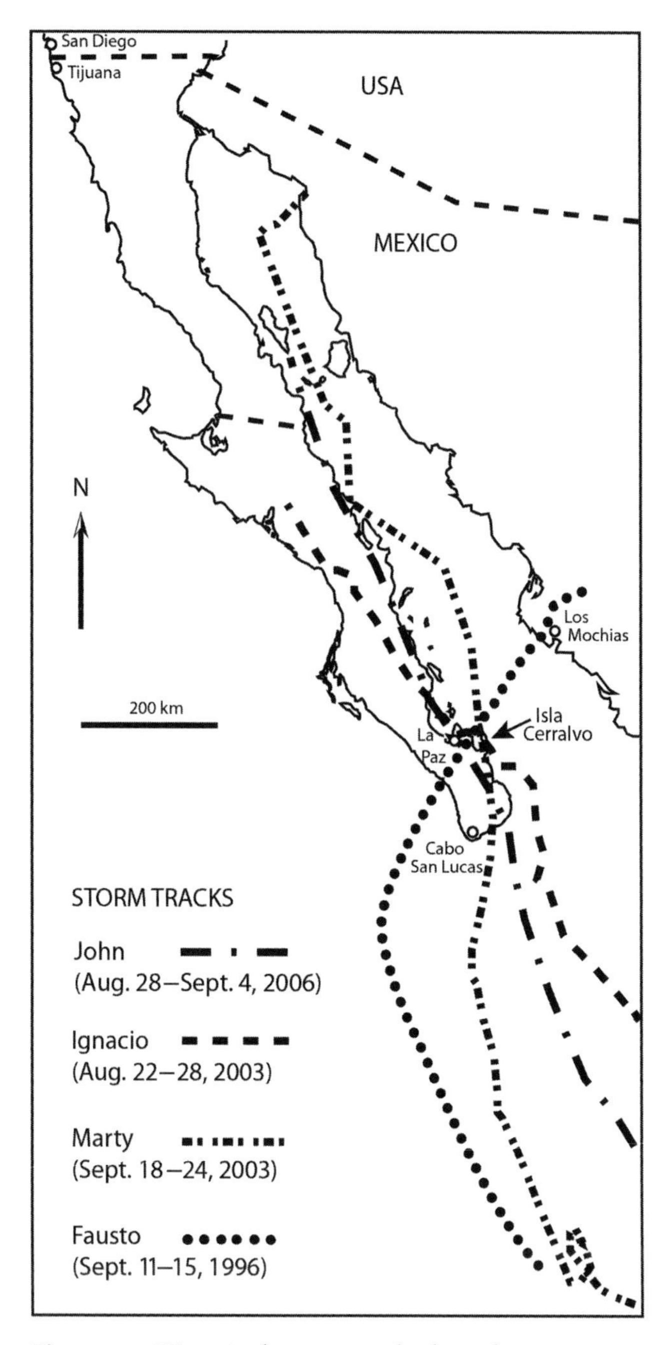

Figure 39. Historical storm tracks from four recent hurricanes striking the Baja California peninsula and Gulf of California. Original drawing by author.

10

Zen Aesthetics and the Big Picture

An Epilogue

In the sea of life, in the sea of death, my soul, tired of both,
seeks the mountain, from which the waters have receded.

—*Unknown author, Japan (seventh century* CE*)*

PROFESSORS GO AWAY ON sabbaticals, surely one of the most generous allowances of any occupation. The leave-takings may even extend to places and topics seemingly unconnected to what the professor normally teaches. In the prelude to an approved absence taking me to East Asia during the 2009 fall term, I found myself standing in line to purchase a ticket to enter part of the Nanzen-ji complex (Southern Zen Temple) on the edge of Kyoto, Japan. I had no legitimate business there but was on my way to do geological fieldwork in South Korea and China. The opportunity to visit dry-landscape rock gardens affiliated with the most famous Buddhist temples in Japan was an experience I could not pass up.

In its own purity, the Gulf of California and the landscapes that enfold it together with all islands, therein, bear the imprint of Zen aesthetics as a way to see the greater world. Marine ecologist Edward Ricketts (1897–1948) and writer John Steinbeck (1902–1968) forged this connection in 1940 when they made a much-celebrated trip to the Gulf of California aboard the chartered fishing vessel *Western Flyer*. The resulting narrative, *The Log from the Sea of Cortez*, is usually seen as an amalgamation of the biological catalogue kept by Ricketts and the prose of Steinbeck based on the trip's journals. The log's most probing entry is for March 24, 1940, often cited as the Easter Sunday sermon. A lucid concluding statement from an otherwise dense manifesto on "non-teleological thinking" offers the following viewpoint (Steinbeck 1951, 150–51): "The whole picture is portrayed by *is*, the deepest word of deep ultimate reality, not shallow or partial as reasons are, but deeper and participating, possibly encompassing the Oriental concept of *being*." Papers edited by Katherine Rodger show that the precise words come from Ricketts, as does much of the entry's previous commentary.[1] Ricketts freely circulated drafts of his essays among close friends, including mythologist Joseph Campbell (1904–1987), but the influence of Oriental philosophy was something the biologist first discovered strictly on his own.[2]

Participating in quite another expedition to the Gulf of California on the research schooner *E. W. Scripps* later the same year in 1940, geologists Charles A. Anderson (1902–1990), J. Wyatt Durham (1907–1996), and Francis P. Shepard (1897–1985) never gained the same level of fame outside their professional circle as did Ricketts and Steinbeck. Of course, insights from the geosciences brought a different perspective to the region's landscape narrative.

Tourists visiting Kyoto temples are not allowed much leisure to contemplate the artfully cloistered and carefully guarded treasures. After purchase of a ticket, you are entrained on a prescribed path and pushed forward to the main features by the press of visitors from behind. There is little time for deep contemplation. My moment of clarity arrived on reaching a particular *nakaniwa*, or enclosed rock garden, in the Southern Zen Temple. I tried to remain on the veranda overlooking the scene as long as possible. Rectangular in layout, the garden consisted solely of several dark stones exhibited on a bed of white, raked gravel (figure 40). For me, it was symbolic of the Gulf of California viewed all at once. One stone was high and peaked like a volcano, others were more medium in height but blocky, while another stood flush with the gravel. Each island stone featured its own set of surrounding ocean ripples inscribed by the raked gravel. Interactions of the Earth's lithosphere, hydrosphere, and atmosphere

Figure 40. Courtyard rock garden from the Nanzen-ji complex (Southern Zen Temple) in Kyoto, Japan. Photo by author.

were neatly tied together in a single unified package. Surely implied but invisible was the Earth's vaporous biosphere.

In practice, Zen aesthetics are not merely visual but entail long periods of seated meditation—the *zazen*. I admit that my meditations were never formal but occurred catch-as-catch-can atop genuine islands in a real sea buffeted by actual winds and currents. My penchant to ask questions on the meaning of things also made me a poor candidate for the Buddhist catechism. There was, I understood, an additional outer envelope to the Earth system not made entirely obvious in the rock garden except for its construction and maintenance by human hands.

French paleontologist and Jesuit priest Pierre Teilhard de Chardin (1881–1955) refined the concept of the noosphere as the "thinking layer" of the Earth generated by progressively mindful beings.[3]

During my brief refuge at the temple *karesansui*, I imagined a web of human relationships cast like an expanding net over the integrated land- and seascapes from the Gulf of California. Human knots binding the web included Ricketts, Steinbeck, Anderson, Durham, and Shephard, but also Ray Cannon, Joseph Wood Krutch, Norman Roberts, C. Carew McFall, and R. Gordon Gastil. Anchor points fixing the net securely in place were individuals such as the ranchers who took Steinbeck and Ricketts hunting for mountain sheep above Puerto Escondido, miners such as Dick Daggett from Los Flores, and homesteaders such as don Chico from San Nicolás. There is a distinct past, present, and future to this web of understanding, just as there is a past, present, and future to the actual landscapes.

The laws of nature all agree on this reality: that the was-ness of becoming must unfold before the is-ness of being takes shape.

Notes

Chapter 1. Cataviña: Gateway to a Peninsular Wilderness

1. The peninsula's botany, including its many cacti, is described by Norman C. Roberts in his *Baja California Plant Field Guide* (La Jolla, CA: Natural History Publishing, 1989). The guidebook cover features a giant cardón over 65 feet (20 m) in height.

2. Ibid., 212.

3. A scheme to compare decades-old photographs of boojum trees with new photos taken from the same vantage point in an effort to estimate the plant's rate of growth is described by Joseph Wood Krutch in *The Forgotten Peninsula*, 2nd ed. (Tucson: University of Arizona Press, 1988), 85–88.

4. Roberts, *Baja California Plant Field Guide*, 126.

5. See figure 5d from Jonathan L. Payne and Markes E. Johnson, "Lower Cretaceous Alisitos Formation at Punta San Isidro: Coastal Sedimentation and Volcanism," *Ciencias Marinas* 30 (2004): 372.

6. John Steinbeck describes his encounters with seafarers in native Nayarit canoes in *The Log from the Sea of Cortez* (New York: Viking Compass, 1951), 97–98.

7. Charles Lyell (1797–1875) is celebrated as the author of *Principles of Geology*, a textbook first issued in three volumes (London: John Murray, 1830–1833) that accounts for geological processes in terms of normal, everyday processes. Many subsequent editions were issued throughout the author's lifetime.

8. The Hyakunin Basho on the grounds of the Imperial Palace in Tokyo is briefly described in *Eyewitness Travel Japan* (London: DK Publishers, 2007), 71.

9. See the geological map of the Baja California peninsula (figure 1.3) in chapter 1 of *Atlas of Coastal Ecosystems in the Western Gulf of California*, edited by Markes E. Johnson and Jorge Ledesma-Vázquez (Tucson: University of Arizona Press, 2009).

10. Drilling depth to granite below the Three Virgins volcanic region is documented in a report by A. Lopez-Hernandez and others, titled "Geological and Geophysical Studies at Las Tres Virgenes, Baja California Sur, Mexico," *Transactions of the Geothermal Resources Council* 18 (1994): 275–80.

11. "Reconnaissance Geological Map of the State of Baja California" (1971) was compiled by R. Gordon Gastil, Richard P. Phillips, and Edwin C. Allison in

three plates (A, B, and C) and published as part of *Memoir 140* by the Geological Society of America.

12. The story of the disappeared Farallon Plate is told in the now-classic paper by Tanya Atwater, "Implications of Plate Tectonics for the Cenozoic Tectonic Evolution of Western North America," *Geological Society of America Bulletin* 81 (1970): 3513–36.

13. Thanks go to Prof. Lee Newsom at Pennsylvania State University for keying woody remains likely to belong to the genus *Cordia* in the borage family.

Chapter 2. Investigations on a Guardian Angel

1. The sizes of the 40 largest islands in the Gulf of California are compared in table 2.2 of Anna Luisa Carreño and Javier Helenes, "Geology and Ages of the Islands," in *A New Island Biogeography of the Sea of Cortés*, ed. T. J. Case, M. L. Cody, and E. Ezcurra, 14–40 (Oxford: Oxford University Press, 2002), 24 and 25.

2. A detailed accounting of the 1539–1540 voyage under the command of Francisco de Ulloa is given in Gorden E. Ness, "The Search for Caláfia's Island," in *The Gulf and Peninsular Province of the Californias*, ed. J. Paul Dauphin and Bernard R. T. Simoneit, 1–23, American Association of Petroleum Geologists Memoir 47 (Tulsa, OK, 1991).

3. The history of the San Francisco de Borja Mission is told and illustrated with recent photographs of surviving buildings by Padre James Donald Francez in *The Lost Treasures of Baja California* (Chula Vista, CA: Black Forest Press, 1996), 57–62.

4. J. Steinbeck and E. Ricketts, *Sea of Cortez: A Leisurely Journal of Travel and Research* (New York: Viking Press, 1941), title page.

5. Katharine A. Rodger, ed., *Breaking Through: Essays, Journals, and Travelogues of Edward F. Ricketts* (Berkeley: University of California Press, 2006), 200.

6. E. F. Ricketts, J. Calvin, and J. W. Hedgeth, revised by D. W. Phillips, *Between Pacific Tides*, 5th ed. (Stanford: Stanford University Press, 1985).

7. Rodger, *Breaking Through*, 170.

8. Quotation from Ray Cannon, *The Sea of Cortez* (Menlo Park, CA: Lane Magazine and Book Company, 1966), 84.

9. Ibid., 91.

10. Peacock went to Isla Angel de la Guarda as a kind of memorial following the death of his friend Edward Abbey. From the prologue to Doug Peacock, *Baja!* (Boston: Bulfinch Press, 1991), 35–49.

11. The flying exploits of Ike Russell are told by his many friends in Thomas Bowen, ed., *Backcountry Pilot* (Tucson: University of Arizona Press, 2002).

12. The adventures of Graham MacKintosh in his circumperambulation around the Baja California peninsula are told in *Into a Desert Place: A 3000 Mile Walk Around the Coast of Baja California* (London: Unwin Hyman, 1988). MacKintosh did make it out to Isla Angel de la Guarda from Bahía de los Angeles

in January 2006. Although he hiked over the center of the island, he made no attempt to walk around the circumference. The story of his Angel de la Guarda experience is told in Graham MacKintosh, *Marooned with Very Little Beer* (San Diego, CA: Baja Detour Press, 2008).

13. University of California–Davis, "News and Information" bulletin for March 31, 2000, describes the boating accident in the Canal de Ballenas that took the lives of five scientists on March 27, 2000.

14. Refer to chap. 1, n. 11.

15. See map of paleoislands on the Punta Chivato promontory in Markes E. Johnson, *Discovering the Geology of Baja California: Six Hikes on the Southern Gulf Coast* (Tucson: University of Arizona Press, 2002), 97, fig. 16.

16. Size variations and other details regarding the comparative life history and reproductive features of *Sauromalus* species are summarized in table 9.5 of Ted J. Case, "Reptiles: Ecology," in *A New Island Biogeography of the Sea of Cortés*, ed. T. J. Case, M. L. Cody, and E. Ezcurra, 221–70 (Oxford: Oxford University Press, 2002), 247.

17. The taxonomic divisions within the cyanobacteria (blue-green bacteria) feature two classes (the Coocogoneae and the Hormogoneae), and the attributes of the various orders within these classes are discussed by Lynn Margulis and Karlene Schwartz in *Five Kingdoms*, 2nd ed. (New York: W. L. H. Freeman, 1988), 48–49.

18. Now recognized as a species endemic to Isla Angel de la Guarda, the rattlesnake *Crotalus angelensis* reaches an adult length of up to five feet (1.52 m) and is "very common on gravelly beaches, rocky arroyos and hillsides," according to Stuart Aitchison, *The Desert Islands of Mexico's Sea of Cortez* (Tucson: University of Arizona Press, 2010), 35.

Chapter 3. Great Sand Ramp at Km 41

1. The estimated total area of arid lands around the world is given in table 5 of "Climate," chap. 1 in *Deserts*, by M. C. Stoppato and A. Bini (Toronto, ON: Firefly Books, 2003), 17; the estimated total area of sand dunes is 5 percent, according to D. Thomas, "Sand Seas and Aeolian Bedforms," in *Arid Zone Geomorphology: Process, Form and Change in Drylands*, ed. D. S. G. Thomas, 373–412, 2nd ed. (Chichester, UK: John Wiley and Sons, 1997).

2. A brief accounting of the Los Flores mines and the Daggett family is offered in Greg Niemann, *Baja Legends* (San Diego, CA: Sunbelt Publications, 2002), 165.

3. Roberts, *Baja California Plant Field Guide*, 211.

4. Identification based on personal communication from Mark Dimmitt (Arizona-Sonora Desert Museum, Tucson, Arizona), who adds that one certainly would not want to walk around the habitat of this grass in flip-flops. The burs are very sharp.

Chapter 4. San Francisquito's Ancient Bay

1. John Steinbeck gives his impression of the land around San Francisquito in *The Log from the Sea of Cortez* (New York: Viking Press, 1951), 211.

2. Ricketts established his first business, Pacific Biological Laboratories, on Ocean View Avenue in Pacific Grove, California; Steinbeck had taken up residence in his parents' summer cottage on Eleventh Street in Pacific Grove. How the two met in 1930 and became close friends is told by Eric Enno Tamm in *Beyond the Outer Shores* (New York: Four Walls Eight Windows, 2004).

3. For a description of living rhodolith beds and the physical conditions that favor their productivity, see Diana L. Steller, Rafael Riosmena-Rodríguez, and Michael S. Foster, "Living Rhodolith Bed Ecosystems in the Gulf of California," in *Atlas of Coastal Ecosystems in the Western Gulf of California*, ed. Markes E. Johnson and Jorge Ledesma-Vázquez, 72–82 (Tucson: University of Arizona Press, 2009).

4. The biostratigraphic ranges of index fossils from Pliocene and Pleistocene strata along the gulf shores of the Baja California peninsula are outlined by J. Wyatt Durham in his monograph *Megascopic Paleontology and Marine Stratigraphy*, part 2 of *1940* E. W. Scripps *Cruise to the Gulf of California*, Geological Society of America Memoir 43 (New York: Geological Society of America, 1950), 40–41 and table 9. Durham's scheme is updated with reference to a master stratigraphic section on Isla Carmen in a research paper by James M. Eros, Markes E. Johnson, and David H. Backus, "Rocky Shores and Development of the Pliocene-Pleistocene Arroyo Blanco Basin on Isla Carmen in the Gulf of California, Mexico," *Canadian Journal of Earth Sciences* 43 (2006): 1149–64.

5. Bivalve species belonging to the family Glycymeridae (the bittersweet shells) are described together with their life habits by Richard C. Brusca in *Common Intertidal Invertebrates of the Gulf of California*, 2nd ed. (Tucson: University of Arizona Press, 1980), 133–36.

6. Ibid., 164.

7. Steinbeck, *Log from the Sea of Cortez*, 212.

Chapter 5. Lost Lagoons of Bahía Concepción

1. The geology and paleontology of the Punta Chivato area, including its several "fossil" islands, are treated in Johnson, *Discovering the Geology of Baja California*.

2. The McFall excursion to Bahía Concepción and the Concepción Peninsula was probably one of the last great projects supported by the Belvedere Scientific Fund, endowed by contributions from key officers of the Bechtel Corporation. McFall's map was drafted at a scale of 1:70,000 and is printed as a foldout measuring 26.5 inches (82 cm) by 22.5 inches (71 cm). The map comes with a 25-page text, "Reconnaissance Geology of the Concepcion Bay Area, Baja California, Mexico," *Stanford University Publications, Geological Sciences* 10, no. 5 (1968).

3. A report by Michael S. Foster and others under the title "Living Rhodolith Beds in the Gulf of California and Their Implications for Paleoenvironmental Interpretation" gives the story of El Requeson as a major center for rhodolith banks in Bahía Concepción: This report is in *Pliocene Carbonates and Related Facies Flanking the Gulf of California, Baja California, Mexico,* edited by Markes E. Johnson and Jorge Ledesma-Vázquez, 127–39, Geological Society of America Special Paper 318 (Boulder, CO: Geological Society of America, 1997).

4. Brusca, *Common Intertidal Invertebrates of the Gulf of California,* 164–65.

5. Consult Mark A. Mayall, "A Comparison of Modern and Ancient Embayments: Mary Creek, St. John, U.S. Virgin Islands, and the Rancho Santa Rosalita Basin, Concepcion Peninsula, Baja California Sur, Mexico," unpublished honors thesis, Williams College, 1993 (159 pp.).

6. The stratigraphy of Rattlesnake Ridge is described in full by Johnson and colleagues on p. 71 of "Upper Pliocene Stratigraphy and Depositional Systems: The Peninsula Concepción Basins in Baja California Sur, Mexico," in Johnson and Ledesma-Vázquez, *Pliocene Carbonates and Related Facies,* 57–72.

Chapter 6. Intersection of Fractures at El Mangle

1. The history of Loreto is told in fascinating detail by Ann O'Neil and Don O'Neil in *Loreto, Baja California: First Mission and Capital of Spanish California* (Studio City, CA: Tio Press, 2001).

2. Ibid., 37, 48–50.

3. The "Israelsky wedge" was introduced in an article authored by Merle C. Israelsky under the title "Oscillation Chart," published in the *Bulletin of the American Association of Petroleum Geologists* 33 (1949): 92–98. The term *oscillation* was applied to relative changes in sea level that shift marine deposits back and forth in equilibrium with a retreating or advancing shoreline. A single cycle of advance (marine onlap) and retreat (marine offlap) was hypothetically shown to produce a nested set of "sideways V-shaped" layers that form a wedge. As a petroleum geologist, Israelsky was interested in correlating oil-rich layers in the subsurface of the southern US coastal plain along the Gulf of Mexico. Although rock layers were originally deposited at different water depths depending on distance from shore, Israelsky emphasized that the inflection point in the change from marine onlap to marine offlap signified multiple points in a line of correlation.

4. The standard mineralogy reference book is *Dana's Manual of Mineralogy,* first issued in 1848 by Yale professor James Dwight Dana (1813–1895) and used by generations of geology students in subsequent editions down to the present day. Specific gravity (or relative density) is a number expressed as a ratio between mineral weight and the weight of an equal volume of water at 4°C.

5. The intricacies of radiometric dates and geological time are surveyed in James G. Ogg, Gabi Ogg, and Filix M. Gradstein, *The Concise Geologic Time Scale* (Cambridge: Cambridge University Press, 2008).

6. Fossil plant affinities for opalite samples from El Mangle are reported in Darius E. Mitchell, David H. Backus, and Markes E. Johnson, "Opalized Wood from the Upper Pliocene of Baja California Sur: A Coastal-Plain Deposit on the Gulf of California," *Geological Society of America Abstracts with Program* 33, no. 6 (2001): A-197–A-198. Comparison with living halophytes was based on scanning electron microscope images of fossil twigs showing preservation of an inner pith core, xylem vesicles with possible sieve plates, and an outer sheath of fine phloem structures. Rare plant fibers with possible affinities to cactus wood also are reported.

7. The most detailed compilation, as provided in plates 1 and 2, is in J. Paul Dauphin and Gorden E. Ness, "Bathymetry," in *The Gulf and Peninsular Province of the Californias*, ed. J. Paul Dauphin and Bernd R. T. Simoneit, American Association of Petroleum Geologists Memoir 47 (Tulsa, OK: American Association of Petroleum Geologists, 1991).

Chapter 7. Coral Reef on a Volcano at Isla Coronados

1. Refer to chap. 2, n. 1.

2. With his survey work aboard the research vessel *E. W. Scripps*, Francis P. Shepard was the first to map the bathymetry of the Carmen Passage (chart 6). See F. P. Shepard, *Submarine Topography of the Gulf of California*, part 3 of *1940* E. W. Scripps *Cruise to the Gulf of California*, Geological Society of America Memoir 43 (New York: Geological Society of America, 1950).

3. In a short piece in *The Dredgings* (vol. 50, 2010), Rick Harbo describes the harvest of pen shells from San Ignacio Lagoon on the west coast of Baja California. A single adductor muscle comparable to that of a big scallop can be taken from the largest shells. Shell heaps typically mark the favorite places where this mollusk is collected. One such spot on the gulf coast of Baja California is the sheltered bay at El Requeson in Bahía Concepción.

4. *Porites panamensis* has a growth rate of 3.6 millimeters per year in the more tropical waters off the west coast of Panama, as recorded in H. M. Guzmán and J. Cortés, "Arrecifes Coralinos del Pacífico Oriental Tropical," *Revisión y Perspectivas: Revista Biología Tropical* 41 (1993): 535–57.

5. The 39-foot (12-m) terrace is a widely occurring physical feature along the gulf shores of Baja California, especially around the Concepción Peninsula. A feature known as Cerro el Sombrerito at Mulegé is described as an eroded volcanic plug with a subsequent marine terrace at this elevation. A paper by J. R. Ashby, T. Ku, and J. A. Minch ("Uranium Series Ages of Corals from the Upper Pleistocene Mulege Terrace, Baja California Sur, Mexico," *Geology* 15 [1987]: 139–141) argues that about half the elevation is due to regional tectonics.

Chapter 8. Song of the Amazon on Isla Monserrat

1. The appropriate passage regarding a society of Amazon women living on the island of "California" (from the 1519 novel *The Adventures of Esplanián*, by

the Spanish author Garcia Ordóñez de Montalvo) is cited in O'Neil and O'Neil, *Loreto, Baja California*, 15.

2. The significance of neritid gastropods as dwellers in the high-intertidal zone of southern Baja California is cited by Ann Zwinger in her natural history of the cape region of Baja California. See Zwinger, *A Desert Country Near the Sea* (Tucson: University of Arizona Press, 1983). See especially the artful photo of *Nerites* shells by Herman Zwinger, 246.

3. Differences between plunging rocky shores and shores with attached Type A and Type B platforms are described by C. D. Woodroffe in *Coasts: Form, Process and Evolution* (Cambridge: Cambridge University Press, 2002). The essential variations relate to the abruptness of a rocky shore with respect to the tidal zone.

4. Park permits issued for research allow for the collection of fossils, and the rhodoliths from the high limestone plateau near the center of Isla Monserrat were sent for identification to Prof. Rafael Riosmena-Rodríguez in the Program for Marine Botany at Universidad Autónoma de Baja California in La Paz. The specimens are deposited with the university museum. Under normal circumstances, visitors to the islands are not allowed to disturb or remove anything whatsoever.

5. Consult chapter 5, "The Far North Shore: Ensenada el Muerto" in Johnson, *Discovering the Geology of Baja California*, 91–119. Therein, the dynamics of sandbar formation off shores with heavy surf are discussed. A photo of the worn shell fragments on this particular beach is shown in figure 18 (103).

6. Author Graham MacKintosh recounts the history of scallop divers working off Isla Angel de la Guarda during the early 1970s in his book *Marooned with Very Little Beer* (2008). Chapter 8 ("Gather Seashells While Ye May," 54–64) is based on accounts of former divers that Graham interviewed in Bahía de los Angeles in 2006. At the height of commercial activities on Angel de la Guarda, there were as many as 600 people living in a support camp on the island.

7. In particular, Dr. Ana Luisa Carreño (Instituto de Geología, Universidad Nacional Autónoma de México) was most helpful in identifying key microfossils from the older Pliocene deposit on Isla Monserrat. She provided the information that put some dozen microfossils retrieved from the locality into a tight chronostratigraphic context.

8. The snowball hypothesis of pecten shells incorporated into armored mud balls was presented by Jorge Ledesma-Vázquez in 2007 at the Fourth European Meeting on the Palaeontology and Stratigraphy of Latin America under the title "Armored Mud Balls in Tidal Environments, Pliocene in the Gulf of California" and published in the conference abstracts: *Cuadernos del Museo Geominero* 8 (Madrid: Instituto Geológico y Minero de España, 2007), 235–38.

Chapter 9. Riding Out Ancient Storms on Isla Cerralvo

1. Additional details provided by the Hertlein report show that the benefactor of the 1965 expedition to Isla Cerralvo was Mr. Harry "Bing" Crosby, famous for his 1950s/1960s singing and movie career.

2. With his survey work aboard the research vessel *E. W. Scripps*, Francis P. Shepard was the first to map the bathymetry of the Cerralvo Passage (chart 8). See chap. 7, n. 2.

3. Herodotus described the Nile River and its delta in detail based on his personal experience visiting Egypt in antiquity. It is noteworthy that Herodotus cites Pharonic names for various distributaries of the Nile (Pelusian, Canobic, and Sebennytic) leading to separate mouths on the Mediterranean Sea. The branching distributaries are what give the classic "delta" shape to the huge estuary system. See the translation by David Grene, *The History* (Chicago: University of Chicago Press, 1987), bk. 2, sec. 17, p. 138.

4. A dozen different delta systems are categorized by George Postma in his classification in Postma, "Sea-Level-Related Architectural Trends in Coarse-Grained Delta Complexes," *Sedimentary Geology* 98 (1995): 3–12. Technically, the structures on the north end of Isla Cerralvo can be identified as Type C shoal-water deltas.

5. Prior to about 3.5 million years ago, strong currents moved the larvae of marine organisms, as well as any free-swimming or floating biota, from the Caribbean into the eastern Pacific Ocean through present-day Panama. For documentation, see Anthony G. Coates et al., "Closure of the Isthmus of Panama: The Near-Shore Marine Record of Costa Rica and Western Panama," *Geological Society of America Bulletin* 104 (1992): 814–28.

Chapter 10. Zen Aesthetics and the Big Picture: An Epilogue

1. Rodger, *Breaking Through*. See especially chapter 5, "Essay on Non-teleological Thinking" (119–33), which shows the extent of Rickett's literary contribution to *Log from the Sea of Cortez*.

2. Among Rickett's treasured books was a copy of *Essays in Zen Buddhism* by Daisetz Teitaro Suzuki in an English translation (1933) published in London by Luzac and Company. See related comments by Tamm in *Beyond the Outer Shores*, 109, and by Rodger in *Breaking Through*, 89.

3. De Chardin's concept of the noosphere is given a more contemporary airing in Edward O. Dodson, *The Phenomenon of Man Revisited: A Biological Viewpoint on Teilhard de Chardin* (New York: Columbia University Press, 1984).

Glossary of Geological and Ecological Terms

aa lava. Hawaiian term for basaltic lava flows characterized by rough or jagged surfaces.

agglomerate. Pyroclastic rocks (volcanic) consisting mainly of rounded or subangular fragments more than 12.5 inches (32 mm) in diameter.

andesite. Igneous rocks from surface flows (volcanic lavas and breccias) rich in the minerals plagioclase and lesser amounts of pyroxene, hornblende, and/or biotite. Although a freshly broken surface is generally dark gray in color, weathered surfaces often take on a reddish cast because of oxidation of iron content. The name is derived from the Andes, which is representative of the mountain belt where these rocks occur because of ocean-plate subduction against a continental margin.

angular unconformity. Break in the rock record in which older strata below a prominent erosional surface dip at an angle that is considerably steeper than the younger strata deposited above.

basalt. Igneous rocks of an extrusive origin (flood lavas or submarine pillow lavas) rich in the minerals plagioclase, pyroxine, and often olivine. These rocks typically form on the ocean floor or in continental rifts. They are dark and generally weather dark.

basement rocks. The intrusive igneous and/or metamorphic rocks overlain at depth by sedimentary rocks.

batholith. Stock-shaped or dome-shaped igneous intrusion generally larger than 40 square miles (103 km^2).

bedrock. Solid rock, stratified or unstratified, that occurs under soil, gravel, or any other unconsolidated surface materials.

berm. Well-demarcated deposit of sand or gravel that is level in extent and marks the maximum landward extent of reworking by waves. A beach may have more than one berm, depending on the effect of tidal cycles and storms.

brachiopod. Shelled marine invertebrate (of the phylum Brachiopoda) with two valves ($CaCO_3$) that are typically different from one another in shape but individually bilaterally symmetrical..

carbonate ramp. Gently sloping surface (generally 5° to 10°) that forms as a continuum from shallow to deeper water and consists of carbonate sediments when under active construction, or limestone when cemented in place. The ramp may sit on an unconformity surface eroded from preexisting rocks (sedimentary, igneous, or metamorphic) by coastal and near-shore processes.

clast. Individual fragment of rock (varying sizes), eroded by the action of wind, waves, or running water from a parent source.

coquina. Bedded accumulations of cemented shells that virtually exclude any intermittent sediment.

coral. Marine invertebrates that are solitary or colonial and belong to the phylum Coelenterata. Many species secrete a solid skeleton ($CaCO_3$) and are the principal contributors to the construction of reefs.

coralline red algae. Marine plants that belong to the division Rhodophyta and have the ability to secrete skeletons ($CaCO_3$). The adjective *coralline* is applied to indicate that the algae mimic corals in appearance.

dike. Tabular body of igneous rock injected as a magma through cracks and fissures in a preexisting body of rock.

echinoderm. Marine invertebrates that are solitary in plan, exhibit fivefold symmetry, and belong to the phylum Echinodermata. Many species secrete a shell (test) formed by $CaCO_3$. Common examples include sea urchins, sand dollars, and starfish.

escarpment. Steep rock face showing the termination of stratified rocks that dip somewhat into the outcrop.

extensional stress. Force applied to the earth's crust that results in stretching or pulling in opposite directions.

facies. Sedimentary rocks and fossils that represent contemporaneous variations in a lateral continuum. A common representation relates to facies changes in an onshore-offshore pattern.

fauna. The animals that live in a given area or environment. A faunal list gives the names of those animals (or fossils) found in a given habitat.

geomorphology. Study of physical landforms and the natural processes that lead to their development at the surface of the Earth.

geopetal. Partial filling in a cavity showing a flat surface that was originally level with outside surroundings. May be used to tell whether rocks have been tilted after the sediment filling was lithified.

gneiss. Metamorphic rock with bands of coarse grains that may or may not be folded by compression.

graben. Elongated fault block that is down-thrown with respect to adjacent blocks.

granite. Igneous rock that cooled far underground with large mineral crystals that typically include feldspar, plagioclase, quartz, and biotite.

grus. Product of decomposed (weathered) granite that leaves behind a poor soil rich in quartz grains and plagioclase.

horst. Elongated fault block that is upthrown with respect to adjacent blocks on either side. The effect is to create a crested topography.

igneous rocks. Rocks cooled from molten material either deep within the Earth's crust (intrusive) or at the Earth's surface (extrusive) as a result of volcanic activity.

inlier. Body of older rocks reduced in size by erosion and subsequently encircled by younger sedimentary rocks that form an unconformity against the preexisting rocks.

joint. Linear fracture in rocks.

karst. Range of landforms that develop both on and within terrain dominated by limestone cover because of dissolution of $CaCO_3$ under a humid climate. The name comes from the Karst district on the coast of the Adriatic Sea.

krummholz. Term derived from German (twisted wood) that applies to trees or shrubs contorted from normal upright growth to a bent or even ground-hugging posture because of the deleterious effects of salt, sand, or ice crystals carried by the wind.

limestone. Sedimentary rock consisting of $CaCO_3$ derived primarily from organic remains of marine invertebrates such as corals, mollusks, echinoderms, and coralline algae.

marine terrace. Narrow coastal rim that usually slopes gently seaward and is veneered by a marine deposit. Formation of the terrace is caused by intertidal erosion, and the position of the terrace depends on changes in global sea level with respect to changes in the local or regional elevation of the coastline.

metamorphic rocks. Rocks either sedimentary or igneous in origin that are subsequently altered by heat and pressure because of deep burial in the Earth's crust. Limestone may be altered to marble, for example, and granite may be altered to schist.

Miocene. Geological epoch, roughly 18 million years in duration, that began about 23 million years ago and terminated a little more than

5 million years ago. All those sedimentary and igneous rocks that originated during that interval are said to belong to the Miocene Series.

mollusks. Marine invertebrates that are solitary in plan and belong to the phylum Mollusca. The phylum includes land and sea snails (class Gastropoda), clams (class Bivalvia), as well as squids and the octopus (class Cephalopoda). In particular, the shelled gastropods and bivalves lend themselves to fossilization.

novaculite. Thick accumulation of bedded chert, typically light colored.

original horizontality. Sediment (such as sand, silt, and clay) deposited underwater tends to disperse evenly on the sea floor and form level layers. When sedimentary rock layers are found that are steeply tilted, it is understood that the configuration of the beds was altered because of tectonic forces well after the solidification of the strata.

orthoconglomerate. Sedimentary rocks consisting of cemented clasts (pebbles, cobbles, boulders) with little intervening matrix.

outlier. Isolated bodies of stratified rock detached from the main outcrop because of erosion of the surround area between the outlier and the rest of the outcrop. Outliers typically form buttes or mesas that may be far removed from similar rocks.

pahoehoe lava. Hawaiian term for basaltic lava flows characterized by a smooth, billowy, or ropy surface.

Pleistocene. Short geological epoch, dating from about 2,588,000 years ago and ending about 10,000 years ago, that bridges the prior Pliocene Epoch and the Holocene (Recent). All sedimentary and igneous rocks that originated during that time interval are said to belong to the Pleistocene Series.

Pliocene. Geological epoch, roughly 2.8 million years in duration, that began more than 5 million years ago and terminated about 2,588,000 years ago. All those sedimentary and igneous rocks that originated during that interval are said to belong to the Pliocene Series.

pluton. Intrusive body of igneous rock formed beneath the surface from slow-cooling magma. Granite, diorite, and tonolite are examples.

pluvial lake. Lake that formed during a geological interval when rainfall was locally more abundant than today. Present desert regions in the Northern Hemisphere typically were subjected to higher rates of rainfall during the various Pleistocene glaciations. When desert conditions returned during the interglacials, the old lakebeds and lake terraces were exposed.

radiolarian. Single-celled animal (protist) with an internal solid skeleton formed by SiO_2.

rhodolith. Particular kind of coralline red algae that grows unattached on the seafloor. The rhodolith assumes a spherical shape because of frequent movement with wave and current activity during the lifetime of the alga. The alga may colonize a tiny piece of shell or a rock fragment as large as a pebble, thereafter growing outward in a concentric pattern.

rhyolite. Volcanic rock formed as a surface flow that is chemically the fine-grained equivalent of granite.

rift zone. Region typically linear in demarcation where a continent has begun to break apart or where ocean crust continues to spread apart in opposite directions.

saltation. Hopping movement of sand grains in a dune because of transfer of energy from one grain to another due to impact in relation to wind transport. The same term is also applied to stream pebbles that are carried along in a skipping movement by currents. From the Latin *saltare*, to jump.

scree. Pile of rock waste found at the base of a cliff or a sheet of coarse debris that covers a cliff or steep mountainside.

sedimentary rocks. Rocks formed by the burial and cementation of inorganic sediments such as pebbles, sand, silt, and clay, or the fragments of broken corals and shells that form limestone.

slickenside. Polished and striated (scratched) surface that results from friction on a fault plane.

specific gravity. Ratio of the weight of a given mineral to the weight of the same volume of water. Galena or lead (Pb) has a specific gravity of about 7.5, while quartz (SiO_2) has a specific gravity of only 2.6.

strata. Layered sedimentary rocks.

stratum. Single layer (or bed) in a sequence of layered sedimentary rocks.

strike-slip fault. Fault in which the net slippage is confined to the direction of the fault strike. That is, movement on opposite sides of the fault trace is seen mainly as lateral, as opposed to vertical. Common synonyms are *wrench fault* and *transcurrent fault.*

stromatolite. Simple microbial deposit made by cyanobacteria, typically laminar in organization, originating far back in Precambrian time. The name derives from the Greek *stromat*, meaning to spread out, and *lithos* for stone and the Latin *stroma* for bed.

superposition. In conformity to the concept of original horizontality, the interpretation that the bottom layer in a sequence of stratified rocks is the oldest bed and the top layer is the youngest bed.

thrombolite. Microbial deposit with a clotted morphology that dates back to the Cambrian. The name comes from the Greek *thrombos* for clot and *lithos* for stone. A related word from the medical profession is *thrombosis*, referring to a blood clot.

tombolo. Sandbar made of loose sediment that connects an island with the mainland.

tonolite. Quartz diorite, a deep-seated (plutonic) rock composed of plagioclase, hornblende, and biotite but especially enriched in quartz.

transform fault. Major fracture that runs perpendicular to an ocean ridge and along which strike-slip movement occurs.

trellised drainage. Stream pattern in which tributaries merge at right angles to one another. Such a pattern is usually controlled by faulted or folded bedrock.

tuff. Rock formed from volcanic ash and small fragments (usually less than 4 mm or ⅛ in. in diameter) of volcanic rock blasted by an eruption.

tuffaceous. Adjective for sediments more than 50 percent composed of tuff.

unconformity. Surface of erosion that separates two bodies of rock and represents an interval of time during which deposition ceased, some material was removed, and then deposition resumed again. An angular unconformity involves two sets of stratified rocks on opposite sides of the unconformity surface, but other types of unconformities may involve a juncture between sedimentary rocks and igneous or metamorphic rocks.

uniformitarianism. Basic concept that the same physical processes that shaped the Earth throughout geologic time in the past are the same processes we may observe in action today.

Geological and Biological References

Anderson, C. A. 1950. *Geology of Islands and Neighboring Land Areas.* Part 1 of *1940* E. W. Scripps *Cruise to the Gulf of California.* Geological Society of America Memoir 43. New York: Geological Society of America.

Ashby, J. R., T. Ku, and J. A. Minch. 1987. Uranium series ages of corals from the upper Pleistocene Mulege terrace, Baja California Sur, Mexico. *Geology* 15: 139–41.

Atwater, T., 1970. Implications of plate tectonics for the Cenozoic tectonic evolution of western North America. *Geological Society of America Bulletin* 81: 3513–36.

Backus, D. H., and M. E. Johnson. 2009. Sand dunes on peninsular and island shores in the Gulf of California. In *Atlas of Coastal Ecosystems in the Western Gulf of California*, edited by M. E. Johnson and J. Ledesma-Vázquez, 117–33. Tucson: University of Arizona Press.

Backus, D. H., M. E. Johnson, and R. Riosmena-Rodríguez. 2012. Distribution, sediment source, and coastal erosion of fan-delta systems on Isla Cerralvo (Lower Gulf of California, Mexico). *Journal of Coastal Research* 28: 210–24.

Bigioggero, B., G. Capaldi, S. Chiesa, A. Montrasio, L. Vezzoli, and A. Zanchi. 1988. Post-subduction magmatism in the Gulf of California: The Isla Coronado (Baja California Sur, Mexico). *Instituto Lombardo (Rendiconti Scienze)* 121: 117–32.

Bigioggero, B., S. Chiesa, A. Zanchi, A. Montrasio, and L. Vezzoli. 1995. The Cerro Mencenares volcanic center, Baja California Sur: Source and tectonic control on postsubduction magmatism within the gulf rift. *Geological Society of America Bulletin* 107: 1108–22.

Brown, E. H., and W. C. McClelland. 2000. Pluton emplacement by sheeting and vertical ballooning in part of the southeast Coast Plutonic Complex, British Columbia. *Geological Society of America Bulletin* 112: 708–19.

Brusca, R. C. 1980. *Common Intertidal Invertebrates of the Gulf of California.* 2nd ed. Tucson: University of Arizona Press.

Cannon, R. 1966. *The Sea of Cortez.* Menlo Park, CA: Lane Magazine and Book Company.

Carranza-Edwards, A., G. Bocanegra-Garcia, L. Rosales-Hoz, and L. P. Galán. 1998. Beach sands from Baja California peninsula, Mexico. *Sedimentary Geology* 119: 263–74.

Carreño, A. L., and J. Helenes. 2002. Geology and ages of the islands. In *A New Island Biogeography of the Sea of Cortés*, edited by T. J. Case, M. L. Cody, and E. Ezcurra, 14–40. Oxford: Oxford University Press.

Carreño, A. L., and J. T. Smith. 2007. Stratigraphy and Correlation for the Ancient Gulf of California and Baja California Peninsula, Mexico. *Bulletins of American Paleontology* no. 371.

Case, T. J. 2002. Reptiles: Ecology. In *A New Island Biogeography of the Sea of Cortés*, edited by T. J. Case, M. L. Cody, and E. Ezcurra, 221–70. Oxford: Oxford University Press.

Dauphin, J. P., and G. E. Ness. 1991. Bathymetry of the Gulf and Peninsula Province of the Californias. In *The Gulf and Peninsular Province of the Californias*, edited by J. P. Dauphin and B. R. T. Simoneit, 21–23. American Association of Petroleum Geologists Memoir 47. Tulsa, OK: American Association of Petroleum Geologists.

Des Marais, D. J. 2003. Biogeochemistry of hypersaline microbial mats illustrates the dynamics of modern microbial ecosystems and the early evolution of the biosphere. *Biological Bulletin* 204: 160–67.

Dorsey, R. J., P. J. Umhoefer, and P. R. Renne. 1995. Rapid subsidence and stacked Gilbert-type fan deltas, Pliocene Loreto basin, Baja California Sur, Mexico. *Sedimentary Geology* 98: 181–204.

Durham, J. W. 1947. *Corals from the Gulf of California and the North Pacific Coast of America*. Geological Society of America Memoir 20. New York: Geological Society of America.

———. 1950. *Megascopic Paleontology and Marine Stratigraphy*. Part 2 of *1940* E. W. Scripps *Cruise to the Gulf of California*. Geological Society of America Memoir 43. New York: Geological Society of America.

Emhoff, K. F., M. E. Johnson, D. H. Backus, and J. Ledesma-Vázquez. 2012. Pliocene stratigraphy at Paredones Blancos: Significance of a massive crushed-rhodolith deposit on Isla Cerralvo, Baja California Sur (Mexico). *Journal of Coastal Research* 28: 234–43.

Forrest, M. J., J. Ledesma-Vázquez, W. Ussler, J. T. Kulongoski, D. R. Hilton, and H. G. Greene. 2005. Gas geochemistry of a shallow submarine hydrothermal vent associated with El Requezón fault zone in Bahía Concepción, Baja California Sur, México. *Chemical Geology* 224: 82–95.

Foster, M. S., R. Riosmena-Rodríguez, D. Steller, and W. J. Woelkerling. 1997. Living rhodolith beds in the Gulf of California and their implications for paleoenvironmental interpretation. In *Pliocene Carbonates and Related Facies Flanking the Gulf of California, Baja California, Mexico*, edited by M. E. Johnson and J. Ledesma-Vázquez, 57–72. Geological Society of America Special Paper 318. Boulder, CO: Geological Society of America.

Fraiser, M. L., and D. J. Bottjer. 2007. When bivalves took over the world. *Paleobiology* 33: 397–413.

Gastil, R.G., R. P. Phillips, and E. C. Alison. 1973. Reconnaissance geology of the 1971 Reconnaissance Geological Map of the State of Baja California [in three sheets].

Gugger, M., S. Lenoir, C. Berger, A. Ledreux, J. Druarat, J. Humbert, C. Guette, and C. Bernard. 2005. First report in a river in France of the benthic cyanobacterium *Phormidium favosum* producing anatoxin-a associated with dog neurotoxicosis. *Toxicon* 45: 919–28.

Hayes, M. L., M. E. Johnson, and W. T. Fox. 1993. Rocky-shore biotic associations and their fossilization potential: Isla Requeson (Baja California Sur, Mexico). *Journal of Coastal Research* 9: 944–57.

Hertlein, L. G. 1966. Pliocene fossils from Rancho el Refugio, Baja California, and Cerralvo Island, Mexico. *Proceedings of the California Academy of Sciences* 30: 265–84.

Horodyski, J. J., and S. P. Vonder Haar. 1975. Recent calcareous stromatolites from Laguna Mormona (Baja California), Mexico. *Journal of Sedimentary Petrology* 45: 894–906.

Johnson, M. E. 1997. Silurian event horizons related to the evolution and ecology of pentamerid brachiopods. In *Paleontological Events: Stratigraphic, Ecological, and Evolutionary Implications*, edited by C. E. Brett and C. Baird, 162–80. New York: Columbia University Press.

———. 2002. *Discovering the Geology of Baja California: Six Hikes on the Southern Gulf Coast.* Tucson: University of Arizona Press.

Johnson, M. E., D. H. Backus, and J. Ledesma-Vázquez. 2003. Offset of Pliocene ramp facies at El Mangle by El Coloradito Fault, Baja California Sur: Implications for transtensional tectonics 407–20. Geological Society of America Special Paper 374. Boulder, CO: Geological Society of America.

Johnson, M. E., D. H. Backus, and R. Riosmena-Rodríguez. 2009. Contribution of rhodoliths to the generation of Pliocene-Pleistocene limestone in the Gulf of California. In *Atlas of Coastal Ecosystems in the Western Gulf of California*, edited by M. E. Johnson and J. Ledesma-Vázquez, 83–94. Tucson: University of Arizona Press.

Johnson, M. E., and M. L. Hayes. 1993. Dichotomous facies on a Late Cretaceous rocky island as related to wind and wave patterns (Baja California, Mexico). *Palaios* 8: 385–95.

Johnson, M. E., and J. Ledesma-Vázquez. 1999. Biological zonation on a rocky-shore boulder deposit: Upper Pleistocene Bahia San Antonio (Baja California Sur, Mexico). *Palaios* 14: 569–84.

Johnson, M. E., and J. Ledesma-Vázquez, eds. 2009. *Atlas of Coastal Ecosystems in the Western Gulf of California.* Tucson: University of Arizona Press.

Johnson, M. E., J. Ledesma-Vázquez, D. H. Backus, and M. R. González. 2012. Lagoon microbialites on Isla Angel de la Guarda and associated peninsular shores, Gulf of California (Mexico). *Sedimentary Geology* 263–64: 76–84.

Johnson, M. E., J. Ledesma-Vázquez, M. A. Mayall, and J. Minch. 1997. Upper Pliocene stratigraphy and depositional systems: The Peninsula Concepción basins in Baja California Sur, Mexico. In *Pliocene Carbonates and Related Facies Flanking the Gulf of California, Baja California, Mexico*, edited by M. E. Johnson and J. Ledesma-Vázquez, 57–72. Geological Society of America Special Paper 318. Boulder, CO: Geological Society of America.

Johnson, M. E., J. Ledesma-Vázquez, and A. Y. Montiel-Boehringer. 2009. Growth of Pliocene-Pleistocene clam banks (Mollusca, Bivalvia) and related tectonic constraints in the Gulf of California. In *Atlas of Coastal Ecosystems in the Western Gulf of California*, edited by M. E. Johnson and J. Ledesma-Vázquez, 104–16. Tucson: University of Arizona Press.

Johnson, M. E., R. A. López-Pérez, C. R. Ransom, and J. Ledesma-Vázquez. 2007. Late Pleistocene coral-reef development on Isla Coronados, Gulf of California. *Ciencias Marinas* 33: 105–20.

Johnson, M. E., D. M. Perez, and B. G. Baarli. 2012. Rhodolith stranding event on a Pliocene rocky shore from Isla Cerralvo in the Lower Gulf of California (Mexico). *Journal of Coastal Research* 28: 225–33.

Krutch, J. W. 1998. *The Forgotten Peninsula: A Naturalist in Baja California*. 2nd ed. Tucson: University of Arizona Press.

Krutch, J. W., and E. Porter. 1967. *Baja California and the Geography of Hope*. San Francisco: Sierra Club.

Landcaster, N., and V. P. Tchakerian. 2003. Late Quaternary eolian dynamics, Mojave Desert, California, 231–49. Geological Society of America Special Paper 368. Boulder, CO: Geological Society of America.

Ledesma-Vázquez, J., R. W. Berry, M. E. Johnson, and S. Gutiérrez-Sanchez. 1997. El Mono chert: A shallow-water chert from the Pliocene Infierno Formation, Baja California Sur, Mexico. In *Pliocene Carbonates and Related Facies Flanking the Gulf of California, Baja California, Mexico*, edited by M. E. Johnson and J. Ledesma-Vázquez, 73–81. Geological Society of America Special Paper 318. Boulder, CO: Geological Society of America.

Ledesma-Vázquez, J., M. E. Johnson, D. H. Backus, and C. Mirabel-Davila. 2007. Coastal evolution from transgressive barrier deposit to marine terrace on Isla Coronados, Baja California Sur, Mexico. *Ciencias Marinas* 33: 335–51.

Ledesma-Vázquez, J., M. E. Johnson, O. Gonzalez-Yajimovich, and E. Santamaría-del-Angel. 2009. Gulf of California geography, geological origins, oceanography, and sedimentation patterns. In *Atlas of Coastal Ecosystems in the Western Gulf of California*, edited by M. E. Johnson and J. Ledesma-Vázquez, 1–10. Tucson: University of Arizona Press.

Logan, B. W. 1961. Cryptozoon and associated stromatolites from the Recent, Shark Bay, Western Australia. *Journal of Geology* 69: 517–33.

Lopez-Hernandez, A., et al. 1994. Geological and geophysical studies at Las Tres Vergenes, Baja California Sur, Mexico. *Transactions of the Geothermal Resources Council* 18: 275–80.

López-Pérez, R. A. 2012. Late Miocene to Pleistocene reef corals in the Gulf of California. *Bulletins of American Paleontology*, no. 383.

MacKintosh, G. 1988. *Into a Desert Place: A 3000 Mile Walk Around the Coast of Baja California*. London: Unwin Hyman.

Margulis, L. 1984. *Early Life*. Boston, MA: Science Books International.

Marrack, E. C. 1999. The relationship between water motion and living rhodolith beds in the southwestern Gulf of California, Mexico. *Palaios* 14: 159–71.

McFall, C. C. 1968. Reconnaissance geology of the Concepcion Bay area, Baja California, Mexico. *Stanford University Publications, Geological Sciences* 10, no. 5.

McLean, H. 1989. Reconnaissance geology of a Pliocene marine embayment near Loreto, Baja California Sur, Mexico. In *Geologic Studies in Baja California*, 17–25. Los Angeles, CA: Pacific Section, Society of Economic Paleontologists and Mineralogists.

Ness, G. E. 1991. The search for Caláfia's Island. In *The Gulf and Peninsular Province of the Californias*, edited by J. P. Dauphin and B. R. T. Simoneit, 1–23. American Association of Petroleum Geologists Memoir 47. Tulsa, OK: American Association of Petroleum Geologists.

Paris, R., F. Lavigne, P. Wassime, and J. Sartohadi. 2007. Coastal sedimentation associated with the December 26, 2004 tsunami in Lhok Nga, west Banbda Aceh (Sumatra, Indonesia). *Marine Geology* 238: 93–106.

Payne, J. L., M. E. Johnson, and J. Ledesma-Vázquez. 2004. Lower Cretaceous Alisitos Formation at Punta San Isidro: Coastal sedimentation and volcanism. *Ciencias Marinas* 30: 365–80.

Perry, C. T. 2005. Morphology and occurrence of rhodoliths in siliciclastic, intertidal environments from a high latitude reef setting, southern Mozambique. *Coral Reefs* 24: 201–7.

Peter, J. M., and S. D. Scott. 1991. Hydrothermal mineralization in the Guaymas Basin, Gulf of California. In *The Gulf and Peninsular Province of the Californias*, edited by J. P. Dauphin and B. R. T. Simoneit, 721–41. American Association of Petroleum Geologists Memoir 47. Tulsa, OK: American Association of Petroleum Geologists.

Postma, G. 1995. Sea-level-related architectural trends in coarse-grained delta complexes. *Sedimentary Geology* 98: 3–12.

Roberts, N. C. 1989. *Baja California Plant Field Guide*. La Jolla, CA: Natural History Publishing.

Rowley, G. D. 1978. Fossil cacti—fact or fable. *National Cactus and Succulent Journal* 33: 19.

Russell, P., and M. E. Johnson. 2000. Influence of seasonal winds on coastal carbonate dunes from the Recent and Plio-Pleistocene at Punta Chivato (Baja California Sur, Mexico). *Journal of Coastal Research* 16: 709–23.

Sankey, J. T., T. R. Van Devender, and W. H. Clark. 2001. Late Holocene plants, Cataviña, Baja California. *Southwestern Naturalist* 46: 1–7.

Santos, A., E. Mayoral, B. G. Baarli, C. M. da Silva, M. Cachão, and M. E. Johnson. 2012. Symbiotic association of a pyrgomatid barnacle and a coral from a volcanic middle Miocene shoreline (Porto Santo, Madeira Archipelago, Portugal). *Palaeontology* 55: 173–82.

Sedlock, R. L., F. Ortega-Gutiériz, and R. C. Speed. 1993. *Tectonostratigraphic Terranes and Tectonic Evolution of Mexico.* Geological Society of America Special Paper 278. Boulder, CO: Geological Society of America.

Sewell, A. A., M. E. Johnson, D. H. Backus, and J. Ledesma-Vázquez. 2007. Rhodolith detritus impounded by a coastal dune on Isla Coronados, Gulf of California. *Ciencias Marinas* 33: 483–94.

Shepard, F. P. 1950. *Submarine Topography of the Gulf of California.* Part 3 of *1940* E. W. Scripps *Cruise to the Gulf of California.* Geological Society of America Memoir 43. New York: Geological Society of America.

Smits, A. W. 1985. Behavior and dietary responses to aridity in the chuckwalla, *Sauromalus hispidus. Journal of Herpetology* 19: 441–49.

Stanley, S. M. 1968. Post-Paleozoic adaptive radiation of infaunal bivalve mollusks: A consequence of mantle fusion and siphon formation. *Journal of Paleontology* 42: 214–29.

Steinbeck, J. 1951. *The Log from the Sea of Cortez.* New York: Viking Compass.

Steller, D. L., R. Riosmena-Rodríguez, and M. S. Foster. 2009. Living rhodolith bed ecosystems in the Gulf of California. In *Atlas of Coastal Ecosystems in the Western Gulf of California*, edited by M. E. Johnson and J. Ledesma-Vázquez, 72–82. Tucson: University of Arizona Press.

Thomas, D. S. G. 1997. Sand seas and aeolian bedforms. In *Arid Zone Geomorphology: Process, Form and Change in Drylands*, edited by D. S. G. Thomas, 373–412. Chichester, UK: John Wiley and Sons.

Tierney, P. W., and M. E. Johnson. 2012. Stabilization role of crustose coralline algae during Late Pleistocene reef development on Isla Cerralvo, Baja California Sur (Mexico). *Journal of Coastal Research* 28: 244–54.

Umhoefer, P. J., R. J. Dorsey, and P. Renne. 1994. Tectonics of the Pliocene Loreto basin, Baja California Sur, Mexico and evolution of the Gulf of California. *Geology* 22: 649–52.

Zanchi, A. 1994. The opening of the Gulf of California near Loreto, Baja California, México: From basin and range extension to transtensional tectonics. *Journal of Structural Geology* 16: 1619–39.

Zwinger, A. 1983. *A Desert Country Near the Sea.* Tucson: University of Arizona Press.

Index

Page numbers in italics refer to illustrations. Page numbers in bold refer to glossary definitions.

About the Author

Markes E. Johnson is the Charles L. MacMillan Professor of Natural Science, Emeritus, at Williams College in Williamstown, Massachusetts, where he taught courses in historical geology, paleontology, and stratigraphy in the Geosciences Department over a thirty-five-year career. His undergraduate education in geology concluded with a BA degree (1971) from the University of Iowa, and his advanced training in paleoecology culminated with a PhD degree (1977) through the Department of Geophysical Sciences at the University of Chicago. With twenty-five years of field experience in Baja California, Johnson has been a semiannual visitor to the frontier states of Mexico, where he habitually led field courses and supervised thesis projects for students from Williams College. He is an authority on the geology of ancient shorelines and the evolution of intertidal life through geologic time based on studies conducted around the world from Western Australia to China's Inner Mongolia to the fringe of Arctic lands across Siberia, Norway, and Canada, as well as comparatively young island groups such as the Seychelles in the Indian Ocean and the Cape Verdes in the North Atlantic. Whether on explorations near or far away, this traveler has always been drawn back to the wild islands in the western Gulf of California and their associated peninsular shores. The author lives with his spouse, Gudveig Baarli, in Williamstown, Massachusetts, where they maintain an active and mutually supportive schedule of ongoing research and writing projects. Current interests and journal publications relate to the travels of Charles Darwin and his formative growth as a field geologist during the voyage of the HMS *Beagle*.